# From Chemistry to Consciousness

Harald Atmanspacher · Ulrich Müller-Herold
Editors

# From Chemistry
# to Consciousness

The Legacy of Hans Primas

 Springer

*Editors*
Harald Atmanspacher
Collegium Helveticum
ETH Zurich
Zürich
Switzerland

Ulrich Müller-Herold
CHN E 24
ETH Zentrum
Zürich
Switzerland

ISBN 978-3-319-82859-6          ISBN 978-3-319-43573-2    (eBook)
DOI 10.1007/978-3-319-43573-2

Printed on acid-free paper

This Springer imprint is published by Springer Nature
The registered company is Springer International Publishing AG Switzerland

# Contents

# Contributors

**Harald Atmanspacher** Collegium Helveticum, University and ETH Zurich, Zürich, Switzerland

**Robert C. Bishop** Physics Department, Wheaton College, Wheaton, IL, USA

**Geoffrey Bodenhausen** Départment de Chimie, Ecole Normale Supérieure, Paris, France

**Richard Ernst** Laboratory of Physical Chemistry, ETH Zurich, Zürich, Switzerland

**Peter beim Graben** Bernstein Center for Computational Neuroscience, Humboldt University, Berlin, Germany

**Domenico Giulini** Institute for Theoretical Physics, University of Hannover, Hannover, Germany; Center for Applied Space Technology and Microgravity, University of Bremen, Bremen, Germany

**Basil J. Hiley** Physics Department, University College, London, UK; Birkbeck College, London, UK

**Ulrich Müller-Herold** Department of Environmental Systems Science, ETH Zurich, Zürich, Switzerland

**William Seager** Department of Philosophy, University of Toronto, Scarborough, ON, Canada

# Introduction

The legacy of Hans Primas is an intellectual invitation to questions, concepts, and working tools that were developed over more than 50 years in the life of a scientist who combined engineering, experimental, theoretical, and philosophical elements in an original and often surprising way. Beginning with spectroscopic engineering and modern theories of molecular matter, he proceeded to basic questions of quantum theory and the philosophy of physics, suggested innovative ways of addressing interlevel relations in the philosophy of science, and introduced cutting-edge approaches in the flourishing young field of scientific studies of the mind—in short, from chemistry to consciousness. It is the purpose of this book to keep this legacy alive and give an impression of its scope. It is not meant to be an adulation. What Primas always appreciated was competent and outspoken criticism.

Born in 1928, Hans Primas was a Professor in the Chemistry Department at ETH Zurich from 1962 to 1995, and maintained his research activities until his death in October 2014. His way of thinking was strongly influenced by research directions in various fields of science: first, the rise of mathematical physics in the twentieth century, requiring that physical theorems must comply with mathematical standards and that formal assumptions must be physically motivated; second, the guiding idea of high-energy physics, according to which symmetries are more fundamental than particles or fields and that basic insights often appear in group-theoretical terms; and third, the area of quantum logics, most notably the concept of partial Boolean algebras, bridging a long-standing gap between science and philosophy.

Although much of Primas' early activities can be seen as "problem solving", his overall approach was directed toward the production of systematic and coherently organized knowledge. Therefore, his sympathy for "research programs" is not surprising—even though he did not formulate his own one in a self-contained manner. A research program consists of a core of undisputed assumptions and a periphery of open problems that can be investigated based on those assumptions. This framework of doing science allowed Primas to meet old challenges and explore new ideas without losing broadly accepted solid ground.

Introducing the notion of classical observables in theoretical chemistry, Primas paved the way to a fresh philosophical perspective on quantum theory, which he formulated in an algebraic framework that was often considered too abstract to yield intuitive insight. Primas disproved this prejudice: although the status of ontic states and observables is by definition outside the domain of engineering and experimentation, it can be consistently related to these epistemic domains. In this way, ontic and epistemic viewpoints, such as Einstein's "realist" and Bohr's "anti-realist" stance, can be seen as complementing rather than contradicting each other. A number of recent approaches share this "metaphysical turn" in the philosophy of physics, overcoming a long time of metaphysical abstinence in large parts of twentieth-century science.

In a similar vein, Primas was among the first scientists who critically and systematically assessed the dogma of reductionism, with the result that many of the claims in the philosophy of science of the 1960s were found to be too superficial, not informed enough, or even plainly wrong. He inspired research to explore how refined concepts of emergence are more appropriate for relations between observables at different levels of description, most famously between mechanics and thermodynamics, and most notably between neuroscience and psychology.

The latest work of Primas moved into issues that are currently located in cognitive science and the philosophy of mind, but whose history covers millennia: the problem of how our minds are related to the material world. Again, he distanced himself from traditional (Cartesian) dualist, materialist or idealist accounts and developed a view, stimulated by Wolfgang Pauli and Carl Gustav Jung, that is today called dual-aspect monism. On this view, the mental and the material are conceived as dual epistemic aspects of one underlying ontic reality that is ultimately undivided. This perspective has picked up remarkable momentum in contemporary mind–matter research and consciousness studies.

Beyond technical skills, research projects of this kind require a great deal of inner freedom and an emphatic independence of transitory "mainstream opinions", scientific "fashion waves", or "old boy networks". In the end, this avantgardist attitude entailed that his publications have been far more read than quoted. However, his influence in the scientific and philosophical communities to which he contributed can hardly be overestimated, exactly because his voice was known to be largely impartial. This is another important piece of his legacy: a style of incorruptibility and a commitment to truth that made him invulnerable against the seductions and temptations of the celebrity shows and business maneuvers of much of contemporary science. And this style could be contagious.

Richard Ernst (Zurich) and Geoffrey Bodenhausen (Lausanne/Paris) address Primas' pioneering work in nuclear magnetic resonance. Ernst, one of Primas' early Ph.D. students, was later distinguished with the first of several Nobel Prizes devoted to this field, ultimately deriving from Primas' early work. Bodenhausen, a former student with Primas and collaborator of Ernst, reports some of the more recent developments to which he has contributed.

Ulrich Müller-Herold (Zurich) was a member of the Primas group from the 1970s up to Primas' retirement in 1995. Much of their work at that time was concerned with a key topic of theoretical chemistry: the apparent conflict between classical observables, such as chirality, and their foundation in quantum mechanics.

Spiced with personal memories, Müller-Herold sketches the history of this line of research and how it developed into later work on broader perspectives, for instance aiming at a better understanding of relations between descriptive levels in the philosophy of science.

Domenico Giulini (Hannover/Bremen) discusses the notion of superselection rules, a concept that is especially relevant for framing the notion of classical observables in algebraic quantum theory and quantum field theory. Giulini's account is inspired by an influential meeting of the decoherence group around Hans-Dieter Zeh, of which he was a member, with Primas in the early 1990s. The contribution addresses some key issues (and key misunderstandings) emphasized by Primas, among them the superposition principle, the notion of dynamical symmetries, and the significance of disjoint states.

William Seager (Toronto) takes up Primas' numerous articles with critical remarks about the traditional position among philosophers of science that chemistry, biology, and even psychology can eventually be reduced to the basic laws of physics. Current work in this field has become far more pluralistic, and the leading antagonist of reduction, the concept of emergence, has significantly gained ground. Seager highlights these developments and indicates an interesting relation to the relative-state approach to quantum mechanics by Everett, popularized as the "many-worlds" interpretation. For a while Primas endorsed this approach, but shifted away from it in his later views.

Robert Bishop (Wheaton) and Peter beim Graben (Berlin) present a refined version of emergence, called "contextual emergence", that implements Primas' ideas about interlevel relations in a formally sound and empirically applicable way. Their contribution uses contextual emergence to discuss another one of his interests: the relation between deterministic and stochastic descriptions in science. Based on the insight that neither of these two descriptive modes is in principle more fundamental than the other, they show how determinism can emerge from stochasticity as well as how stochasticity can emerge from determinism. This result obviously impacts the discussion of mental causation and free will.

Basil Hiley (London) was a long-standing collaborator of David Bohm, together with whom he developed an overall picture of mind–matter relations closely related to the dual-aspect monist conjecture due to Pauli and Jung. And, like Primas, they tried to express it in algebraic terms, formulated slightly differently but in a similar spirit. This is the topic of Hiley's contribution. It describes and comments on the parallels between his work on a non-Boolean implicate order and its explicate Boolean projections on the one hand and Primas' ontic-epistemic distinction on the other, where contextual patterns arise from intrinsic structures.

Harald Atmanspacher (Zurich) maintained close contact with Primas over the last 25 years of his life, and worked with him during this time. His contribution outlines the structural relationship between ontic and epistemic descriptions with its implications for concepts such as measurement or emergence. It also presents the current status of another focus of Primas' interest, the application of noncommutative (quantum) structures to fields outside (quantum) physics. And finally there is Primas' innovative idea of transforming the mind–matter problem into the problem of how tensed mental time is related to tenseless physical time. Primas left behind an extensive book manuscript about this topic, which is going to be published soon.

The present collection of essays is based, in part, on a symposium "The Legacy of Hans Primas" at Collegium Helveticum, a transdisciplinary institute jointly operated by ETH Zurich and University Zurich, on November 27, 2015. The complete audio-visual recordings of the full-day symposium are accessible at www. multimedia.ethz.ch/speakers/collegium_helveticum/Einzel veranstaltungen/legacy_primas.

In addition to the chapters stemming from the symposium itself, this volume contains solicited contributions covering further areas of Primas' work and their relation to current research topics. Needless to say, more is yet to be discovered. The complete list of publications by Hans Primas at the end of this volume may serve to engender further ideas and perspectives.

The preparation of the symposium was facilitated by Konrad Osterwalder, rector of ETH at the time of Primas' retirement. Sarah Springman, present rector of ETH, provided substantial financial help to realize the symposium. Gerd Folkers, Director of Collegium Helveticum until early 2016, offered his unrestricted support to make the event possible in the facilities of the Collegium. Finally, Angela Lahee at Springer International Publishing arranged for the volume to be part of Springer's science publication program. We do gratefully appreciate all this encouragement and endorsement.

However, most of our gratitude clearly goes to our colleagues who contributed to this volume. For none of them writing essays such as these is their regular day job. Their dedicated willingness to carry the work of Primas into future directions in science and philosophy cannot be applauded enough. Without their commitment and their reliable and efficient cooperation this volume would not have become reality.

# Hans Primas and His Early Pathway

Richard Ernst

**Abstract** Richard Ernst, Hans Primas' second Ph.D. student, gives a brief overview of Primas' early achievements in the field of nuclear magnetic resonance and of his role in the development of a young branch of science.

## 1 Early Days

It is indeed a great opportunity for me to express my limitless gratitude to Hans Primas. Without his foresight, without his friendship, without his generosity, I would not be here. He has deeply impressed me and changed my attitudes within and beyond the world of science during the past fifty years. He has changed my little world.

Primas was born in Zurich in 1928. His parents were Elizabeth Podnetzki and Jaroslaw Johann Primas. They were responsible for his muse, as it were: they provided everything he ever needed or wanted, like various gadgets that he could play with (Fig. 1). He possessed essentially all sets that were available at that time for experiments: a *Technikus* for mechanical experiments, an *Elektrotechnisches Experimentierbuch* for electrical experiments, an inductor to make flashes, an *Elektromann* set for electrical experiments, a *Radiomann* for building a complete radio, a set for high-frequency experiments, and a set for optical experiments. All these tools were at his disposal for learning and for experimenting. On all books in his possession he put his *ex libris* mark, with a triangle and the text *Hans Primas*. This gives a picture of the world of Hans Primas as a boy.

He frequently visited the Pestalozzi library in Zurich where the 13-year old Hans Primas from Zürich-Höngg struck people as a quiet young man (Fig. 2). He was absorbed by a whole-hearted devotion to research, his spirit being stronger than any external force. What ever he did came from within, from his own will, from the devotion to his goals. According to a report in the Swiss magazine *Schweizer*

Edited by Geoffrey Bodenhausen.

R. Ernst (✉)
Laboratory of Physical Chemistry, ETH Zurich, Zurich, Switzerland
e-mail: ernst@nmr.phys.chem.ethz.ch

© Springer International Publishing Switzerland 2016
H. Atmanspacher and U. Müller-Herold (eds.), *From Chemistry to Consciousness*, DOI 10.1007/978-3-319-43573-2_1

**Fig. 1** Kit for high-frequency experiments, bearing the stamp that Hans Primas used to insert into all books that he read and all kits he used to play with

*Illustrierte* radio technique and chemistry were his world. He did not care about classical literature and fiction, even though there was plenty available in the library. He always picked technical books, seeking to deepen his understanding of the world.

When Primas was fourteen years old he caught a serious case of typhus that prevented him from attending his school, since he had to spend a few months in hospital. He had a lot of time to read, including about cures for the disease that he had caught. To recover, he stayed some time in a sanatorium in Arosa, where he read about X-ray theory and related subjects. He continued to study during this time but missed out on subjects of lesser interest and importance to him. After fully recovering, he could not go back to highschool because he had missed many courses, so he started training for an apprenticeship as a laboratory assistant in the analytical chemistry laboratory of the *Werkzeugmaschinen-Fabrik Oerlikon*. Consequently, he could not pass a regular *matura* (final highschool exam) and had to look for a different, less conventional access to science. For science was his world and he wanted to become part of it.

In this laboratory he carried out down-to-earth experiments. But it was evident that he wanted to pursue an academic education. After completing his apprenticeship, he started studying chemistry at the Winterthur Technical School (*Technikum Winterthur*). Three years later, he completed the course as the best student in his field. He really performed well: whatever he did, he always did it very well. His special talents were recognized by his teachers, especially by Professor Anton Stieger, who became one of his mentors, and whose support would be of great benefit to Primas. Anton Stieger was acquainted with Hans Heinrich Günthard, who was to become Primas' future boss. It was Stieger who convinced Günthard that Primas should be

**Fig. 2** Hans Primas in 1941
(at age 13) in the public
Pestalozzi library in Zurich

admitted to ETH in a rather unusual manner as a special curriculum student, as a so-called *Fachhörer* (Fig. 3), i.e., as a student who cannot be formally enrolled because he did not fulfill the requirements. Primas was expected to pass his entrance examinations at some later time.

Later Primas wrote an interesting text about his unusual route to science[1]:

> During my time as an undergraduate at the Winterthur Technical School, he [Stieger] introduced me to a former student of the *Technikum*, Hans Heinrich Günthard, who was at that time a member of the scientific staff of the Laboratory of Organic Chemistry at ETH. Günthard's advice was to be decisive for me. To his dismay, however, I did not follow up on his most important advice and failed to catch up on my matura, which would have allowed me to get properly enrolled at ETH. Since he insisted so pervasively, I looked at a course intended to prepare a matura at the *Minerva* school, but it took me no more than a week to be convinced that I would never survive this kind of course. So I enrolled as a *Fachhörer* in the department of physics and mathematics at ETH. In principle, this left the option of a retroactive recognition of courses I had completed, in case I would manage to pass the matura exams after all. My enrollment gave rise to something of a storm in a teacup. In addition to the regular courses of the first term, I wanted to attend the special lectures for advanced students given by Wolfgang Pauli. I was called to the rector's secretary to be instructed that this was not possible, and when this turned out to be of no avail, I was summoned to the rector himself. Only when, upon my inquiry, the rector admitted that it was not legally forbidden, was I reluctantly allowed to go ahead.

---

[1] This quote is from a private document by Primas, translated by GB.

**Fig. 3** Authorization delivered by the ETH administration to the effect that Primas could attend lectures as "Fachhörer" (listener) without fulfilling the formal requirements to enroll as a regular student

In fact, Primas would more than once give rise to such reactions in the course of his life—people shaking their heads with incredulous disapproval—but he was always right.

At this time he had already completed two publications that went back to his earlier days as a student at the *Technikum*. "Spot reactions" (Osimitz and Primas 1950) were the subject of his very first published paper. His second publication was concerned with "modern procedures for the qualitative analysis of cations" (Primas et al. 1950). These early papers may not have been crucial for the history of science, but they certainly played a key role in Primas' life.

## 2  Physical Chemistry at ETH

Günthard decided that he should somehow keep Hans Primas busy (Fig. 4). At that time, he was committed to vibrational infrared spectroscopy and thought that Primas should also become proficient in this specialty. And indeed: Primas wrote no less than six papers together with Günthard about the theory of vibrational spectroscopy.

Fig. 4 Hans Heinrich
Günthard brought Primas to
the organic and physical
chemistry laboratories at
ETH

I will not discuss these papers in any detail, save for the fact that concepts of symmetry were recurrent themes. Symmetry was very important for Günthard, and would later become equally important for Primas. Graph theory and molecular orbital theory were among the subjects of these early papers (Günthard and Primas 1956).

But in fact Günthard had a completely different objective in mind. He anticipated that Primas would develop into a specialist of nuclear magnetic resonance (NMR) spectroscopy. At that time, there were no NMR instruments at ETH, but he had the foresight that Primas would earn a reputation in this field. And so Günthard provided the inspiration for him to move into NMR. At that time, Primas knew little about NMR beyond its most basic aspects: that molecules consist of electrons and nuclei, and that a nucleus is not merely a simple sphere but possesses a magnetic moment. If one applies a magnetic field, the magnetic moment starts to rotate in proportion to the field. That is what NMR is all about. And that is how Primas introduced it to me a few years later. It would become the basis of my own work.

Initially, Primas and Günthard decided to use a permanent magnet since it would be sufficiently stable and homogeneous. The magnet was designed and built in-house. In order to demonstrate that theory and engineering are not sufficient and that experiments are also important, Primas carried out a few experiments. Thus, he gave a demonstration of magnetic resonance with a spectrum containing different resonance frequencies that correspond to different nuclear species contained in a molecule. For ethanol, for example there are three lines that tell you what the molecule is. This was the first documented low-resolution spectrum recorded at ETH in April

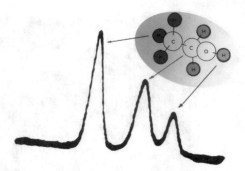

**Fig. 5** Proton spectrum of ethanol with its telling peaks recorded at ETH in 1955 on a spectrometer designed by Hans Primas

**Fig. 6** Hans Primas in conversation with Vladimir Prelog (*right*) in 1986, one of the first organic chemists to realize the potential of NMR for chemistry

1955 at 25 MHz proton resonance frequency (Fig. 5). Soon afterwards, however, it was found that even permanent magnets are not sufficiently stable for demanding high-resolution NMR experiments.

Hans Primas paid careful attention to people who showed interest in NMR, especially Professor Vladimir Prelog,[2] who came by every other day to watch the progress of the NMR spectrometer since he was keen to use it as soon as possible (Fig. 6). So Primas had recruited a first distinguished user for his instrument, long before it was actually finished. Shortly afterwards, some Ph.D. students joined the team of Hans Primas. The first one was Rolf Arndt who finished his Ph.D. in 1962. He was working on solid-state NMR, which offered a completely uncharted territory in those

---

[2]Prelog was a chemist from Croatia, who worked at the Laboratory for Organic Chemistry at ETH Zurich from 1941 to 1976. For his work on stereoisomers he received the Nobel Prize for Chemistry in 1975.

early times, while Primas himself was working on liquid-state resonance. Not much later than Rolf Arndt, it was my turn to join the team.

Primas published two papers that describe the details of his next generation spectrometer. The first one was, as usual, a theoretical paper on lineshape anomalies in high-resolution NMR spectroscopy (Primas 1957). The second, which he published together with Günthard, described how to build an apparatus for high-resolution NMR spectroscopy (Primas and Günthard 1957a). Based on this description, it would actually be possible to build a copy of the apparatus. This important paper represents a major step forward, since nobody in the world had ever built a spectrometer like this before. The assembly of the probe at the heart of the spectrometer is a typical design by Hans Primas (Fig. 7). In this design, symmetry is very important: there is a rotational symmetry when the probe is rotated about its axis during the operation of the spectrometer. So, everything had its meaning.

## HELVETICA PHYSICA ACTA
### Volumen XXX, Fasciculus quartus (1957)

---

### Ein Kernresonanzspektrograph mit hoher Auflösung
#### Teil II: Beschreibung der Apparatur

von H. Primas und Hs. H. Günthard

Organ.-chem. Laboratorium der Eidg. Technischen Hochschule, Zürich.

(27. III. 1957.)

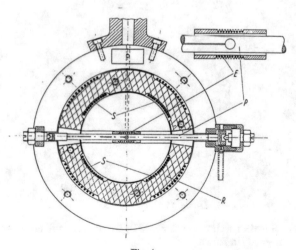

Fig. 4.
Messkopf (geöffnet mit Schnitt durch die Empfänger- und Senderspule) ($E$ = Empfängerspule, $S$ = Senderspule, $R$ = Plexiglasring, $P$ = Probenhalter).

**Fig. 7** Design by Primas of an NMR probe based on symmetry principles

## 3  Commercialization

The results were rather disappointing. I was personally involved in completing the spectrometer, but we soon found out that it was not really useful for recording spectra, since it took too much time. It was really like a snail crawling through a spectrum, recording line by line. So, what should we do with such a beautiful spectrometer if it could not be used? Now, if our wonderful spectrometer failed to deliver, why not commercialize it? There are so many useless objects on the market, one more does not really matter. At that point, Primas knew about a company called Trüb-Täuber ("we build all instruments for measurements"). Initially, the company decided it was an impressive but useless instrument. So Primas and Günthard went to the company atAmpèrestrasse and met Lieni Wegmann who was working on an electron microscope, and asked him: "Mr. Wegmann, can you also build NMR spectrometers?" and he replied: "We will try". So they tried, and built a spectrometer that was basically a copy of Primas' design (Fig. 8).

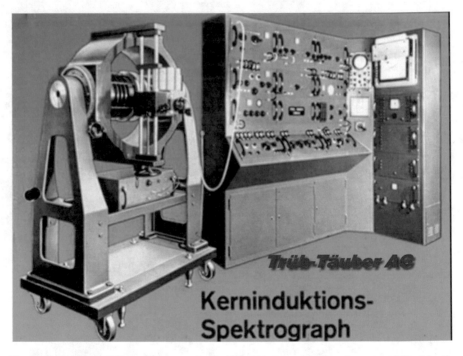

**Fig. 8** Commercial NMR spectrometer designed by Primas and built and commercialized by the company Trüb-Täuber AG in Zurich

**Fig. 9** A list of institutions that purchased one of the early NMR spectrometers designed by Hans Primas and built by Trüb-Täuber AG

- Physikalisches Institut der Universität Genf,
- CNRS Laboratoire de Chimie-Physique, Bordeaux (Prof. Pacault),
- Farbwerke Höchst,
- Karl Marx Universität (Leipzig),
- Shell Grundlagenforschung, Birlinghoven (Prof. Korte) (2 Geräte),
- Phsikalisch-chemisches Laboratorium der Technische Hochschule München (Prof. Scheibe),
- Faculté des Sciences de Clermont-Ferrand, Service Physique (Prof. Raouelt),
- Faculté des Sciences, Paris, Laboratoire de Spectroscopie Hertzienne (Prof. Freymann),
- Universität Tübingen,
- Euramtom CCR, Ispra (Italien),
- Maschpribointorg Moskau (2 Geräte),
- DIA-Berlin,
- Chemische Werke Hüls (Marl),
- Chemisches Insitut der Universität Bonn (Prof. Tscheche),
- Organisch-chemisches Institut der Universität Göttingen (Prof. Glemser),
- Appareils Electriques, Computeurs Garnier (Paris),
- Elektrim, Warschau,
- Univ. Rennes,
- Univ. Liège,
- Prof. Zeil,
- Prof. Brusset (Paris).

They produced as many as 25–30 of these spectrometers and delivered them to various laboratories all over Europe, as listed in Fig. 9.[3] The 25 MHz KR-1 spectrometer had a permanent magnet, while the 75 MHz KR-2 spectrometer with its console was obviously more advanced.[4] The entire development was the brainchild of Hans Primas. Later, these products initiated the development of Bruker as a company.[5] Indeed, the first Bruker spectrometers looked somewhat similar, with a similar magnet and a similar console. This was the beginning of the commercial exploitation of Hans Primas' ideas.

---

[3]Editor's comment: It is the length of the list, more than the people behind the purchases, that impresses modern NMR practictioners. It implies sales of many millions of Swiss Francs. Inter alia, Hans Primas was a successful businessman.

[4]It was a state-of-the-art high-frequency device designed by Hans Primas from scratch; it was part of my thesis to build a probe assembly and a low-noise preamplifier for this instrument.

[5]Trüb-Täuber did not perform very well and the spectrometers did not sell as well as expected, so after all they turned out to be a bit of a flop. Aware of this situation, the German Professor Günther Laukien bought part of the company and continued developing NMR spectrometers, then under the name of "Spectrospin AG".

Of course, the magnetic field of the spectrometer had to be as homogeneous as possible. Together with Günthard, Primas wrote a paper in 1957 showing how to obtain a very homogeneous magnet field (Primas and Günthard 1957b). The field had to be sufficiently stable to record reliable spectra. Further field modulations were needed in order to suppress the background noise. Primas then collaborated with a group gathered around Arndt and myself who had to work out the details. In total we published no less than four papers on the construction of high-resolution NMR spectrometers and problems related to NMR instrumentation.

A brilliant idea came to Primas' mind when he invented the so-called "direct method" (Primas and Günthard 1958). Normally, spectra are rationalized by an indirect method: lines and resonance frequencies are recorded and the transitions are then compared with differences between the eigenvalues of the system. The direct method, on the other hand, allows one to predict the full spectrum directly, without using eigenvalues or eigenfunctions. This was an entirely new method for the analysis of high-resolution NMR spectra.

Together with one of his co-workers, an English postdoctoral fellow, he worked out this direct method in great detail around 1962 and applied it to practical examples. This lead to a very important paper (Banwell and Primas 1963). Primas himself did not enjoy doing "useful" experiments. He much preferred "cute" experiments, and this was a perfect example indeed.

Another great idea by Primas was to use noise for exciting NMR, modeled by a "stochastic" Hamiltonian (Primas 1961). The idea was entirely his, and it would ultimately lead to my Ph.D. thesis (Ernst and Primas 1963). By using stochastic white noise, all resonance frequencies can be excited at once. So the same procedure that was used in Palo Alto by pulsed excitation could be applied in Zürich by stochastic excitation. This opened the way to multiple-frequency excitation, broad-band spectroscopy, multidimensional NMR, higher-order correlation functions, and noise decoupling. All these applications resulted from concepts that Primas had developed. His theoretical work on NMR culminated with a generalized perturbation approach (Primas 1963), which was typical for him as he wanted to be as general as possible—although for the layman the papers are a little bit hard to understand. But the results were impressive.

Before Primas turned away from engineering applications, he had an idea that he gave to his Ph.D. student Adalbert Huber. It was a concept to generate very homogeneous magnetic fields, based on a clever idea (Huber and Primas 1965): "What you have to do is to follow the potential surface of the magnetic field with the surface of your magnet. That allows you to achieve a high-homogeneity magnetic field." This was an engineering concept due to Primas that really worked and turned out to be quite useful (Fig. 10).

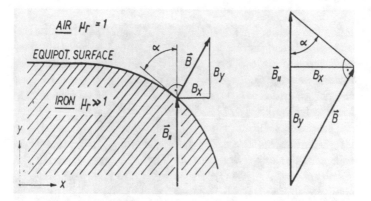

**Fig. 10** Design by Primas for magnet polecaps based on magnetic equipotential surfaces

## 4 California

For me it was time to leave Zurich since I did not have much hope to find a job in Switzerland. I left for California in the United States on a shaky boat. At about the same time Primas also left the field, not physically, but mentally, since he switched from NMR to quantum chemistry. In the end, his reputation would not be built on NMR alone.

In the United States I met my second boss Weston Anderson, who was just as impressive as Hans Primas. He filled some gaps that Primas had left. A significant challenge that Primas had left aside was pulse spectroscopy. He had not worked on the excitation of nuclei by pulses followed by recording of the signals. Anderson invented a "prayer wheel" which he used for exciting all nuclei together, so as to reduce performance time quite drastically.

In California, I became Weston Anderson's "Swiss slave". With a multichannel spectrometer it became possible to record spectra very rapidly. You do not have to sweep the frequencies anymore, as Hans Primas had been doing up to that time. It is a very simple idea, but it changed the world of NMR spectroscopy profoundly. By Fourier transformation one could convert a free induction decay into a spectrum (Fig. 11). This made it much more convenient to record spectra. Primas had overlooked the point, or at least overestimated the difficulties.[6]

Weston Anderson and I filed a patent about this technique at Palo Alto, but nobody was interested in it. So the company "Spectrospin" that had collaborated with Primas built a spectrometer by following the specifications of the patent. It was Tony Keller, a Swiss engineer at Spectrospin, who actually copied the design. This was an important step. Based on these early achievements, the company "Bruker BioSpin" would be established many years later. Today Bruker covers 90 % of the world market in NMR

---

[6]It appears that Primas did not try pulse excitation because it was considered to be too difficult. He wrote about noise excitation but did not actually try to implement it experimentally.

**Fig. 11** Richard Ernst, Primas' second Ph.D. student, developing the first Fourier transform NMR spectrometer in Palo Alto in California, after completing his Ph.D. at ETH

instrumentation, based on the idea of pulsed excitation—an idea that actually made some money.

After my return from the USA in 1968, Hans Primas and I did not meet each other very often any longer, for we had developed different interests in the meantime. Although I would become deeply involved in Buddhist philosophy and Asian religion, this turned out to be insufficient to build bridges to C.G. Jung and Hans Primas. Nevertheless, our parallel interests in science and the arts gave me a feeling of a profound intellectual proximity with Hans Primas. But the time that we could spend together was too short to discuss these matters in any depth, and our later contacts were too limited.

Primas' retirement party took place in 1995 (Fig. 12). Carl Friedrich von Weizsäcker, Peter Pfeifer, and Vittorio Hoesle were the main speakers on this particular occasion, where I entertained the audience by playing an extended single tone ostinato bass on a viola da gamba (Fig. 13). Never before had I held a viola da gamba in my hands, but the audience appeared to be satisfied with my performance. Of course, all this happened long after Primas had left magnetic resonance.

This was just a brief story of Hans Primas' NMR adventures. For a more comprehensive account of Primas' work in NMR see the article by Ernst (1999). His life went on, he moved into quantum chemistry and quantum mechanics in a broader sense. The other presentations in this volume will be concerned with his achievements in these fields.

**Fig. 12** Hans Primas lecturing in the auditorium maximum of ETH on the occasion of the Nobel festivities in 1992 for Richard R. Ernst in 1995

**Fig. 13** Richard Ernst playing the viola da gamba at Primas' retirement party in 1995

# References

Banwell, C.N., and Primas, H. (1963): On the analysis of high-resolution nuclear magnetic resonance spectra. I. Methods of calculating NMR spectra. *Molecular Physics* **6**, 225–256.

Ernst, R.R. (1999): Hans Primas and nuclear magnetic resonance. In *On Quanta, Mind and Matter. Hans Primas in Context*, ed. by H. Atmanspacher *et al.*, Kluwer, Dordrecht, pp. 9–38.

Ernst, R., and Primas, H. (1963): Nuclear magnetic resonance with stochastic high-frequency fields. *Helvetica Physica Acta* **36**, 583–600.

Günthard, Hs.H., and Primas, H. (1956): Zusammenhang von Graphentheorie und MO-Theorie von Molekeln mit Systemen konjugierter Bindungen. *Helvetica Chimica Acta* **39**, 1645–1653.

Huber, A., and Primas, H. (1965): On the design of wide range electromagnets of high homogeneity. *Nuclear Instruments and Methods* **33**, 125–130.

Osimitz, F., and Primas, H. (1950): Tüpfelreaktionen. *Schweizerische Laboranten-Zeitung* **7**, 2–7.

Primas, H. (1957): Ein Kernresonanzspektrograph mit hoher Auflösung. I. Theorie der Liniendeformation in der hochauflösenden Kernresonanzspektroskopie. *Helvetica Physica Acta* **30**, 297–314.

Primas, H. (1961): Über quantenmechanische Systeme mit einem stochastischen Hamiltonoperator. *Helvetica Physica Acta* **34**, 36–57.

Primas, H. (1963): Generalized perturbation theory in operator form. *Reviews of Modern Physics* **35**, 710–712.

Primas, H., and Günthard, Hs.H. (1957a): Ein Kernresonanzspektrograph mit hoher Auflösung. II. Beschreibung der Apparatur. *Helvetica Physica Acta* **30**, 315

Primas, H., and Günthard, Hs.H. (1957b): Herstellung sehr homogener axialsymmetrischer Magnetfelder. *Helvetica Physica Acta* **30**, 331–346.

Primas, H., and Günthard, Hs.H. (1958): Eine Methode zur direkten Berechnung des Spektrums der von quantenmechanischen Systemen absorbierten bzw. emittierten elektromagnetischen Strahlung. *Helvetica Physica Acta* **31**, 413–434.

Primas, H., Lasman, H., and Osimitz, F. (1950): Moderne Vorschriften zur qualitativen Kationenanalyse. *Schweizerische Laboranten-Zeitung* **7**, 98–114.

# Hans Primas—An Inspiring Teacher

Geoffrey Bodenhausen

**Abstract**  The author, who was an undergraduate student at ETH (1970–1974) and came back as a post-doctoral fellow (1980–1985), gives a personal account of Hans Primas' role as a remarkable teacher who triggered fruitful thinking about symmetry and superoperators.

## 1 Introduction

When I was an undergraduate student at ETH, Hans Primas was one of the most impressive of many brilliant faculty. I clearly and dearly remember his lectures on quantum mechanics ("chemische Bindung") and his introduction to group theory. Primas' lectures left a deep impression—not merely of crystalline clarity, but of the powerful idea that it is worth seeking clarity. He repeatedly made the point that the main challenge is to formulate a good question, since it is comparatively trivial to come up with answers. His lectures were punctuated with his favorite expression "salopp gesagt", much more picturesque than feeble translations like "roughly".

At that time, Primas had not yet written his book on group theory (Primas 1978), so he encouraged us to read a textbook by Mathiak and Stingl (1968), which I literally devoured (almost every sentence in my copy of the book is underlined in pencil!). Primas had an unusual ability to stimulate one's curiosity. He had an uncanny gift to bring even boring items such as character tables to life. He has been a source of inspiration and a model ever since—perhaps one of the motivations for me to become a teacher and researcher myself.

Primas was an active member of various faculty committees such as the *Abteilungsrat* and the *Reformkommission* of the ETH in the 1970s. He had a reputation of being open-minded, less concerned with the defense of his professorial status and *Standesinteressen* than many of his touchy colleagues, and genuinely curious to hear if student representatives had anything meaningful to say. (Predictably, we failed to deliver any insights worthy of Primas' expectations. In retrospect, our mag-

G. Bodenhausen (✉)
Départment de Chimie, Ecole Normale Supérieure, Paris, France
e-mail: geoffrey.bodenhausen@ens.fr

H. Atmanspacher and U. Müller-Herold (eds.), *From Chemistry to Consciousness*, DOI 10.1007/978-3-319-43573-2_2

azine "Hundazon" was hardly worth reading, though at the time its editors believed in its revolutionary virtues.) I admired Primas' sense of humor and realism when he spoke in the *Abteilungsrat* and in the *Reformkommission*.

## 2   Symmetry at the Heart of Magnetic Resonance

In the realm of magnetic resonance, Primas' role and influence are evident. Group theory—in particular the classification of quantum states and spectral transitions according to irreducible representations of permutation groups—is a cornerstone of both vibrational spectroscopy and magnetic resonance. It is no accident that Primas worked in both areas in his early days. Some of our early papers (e.g., on methyl groups in proteins; Müller et al. 1987) are largely variations on themes that were first encountered in Primas' lectures (see Fig. 1).

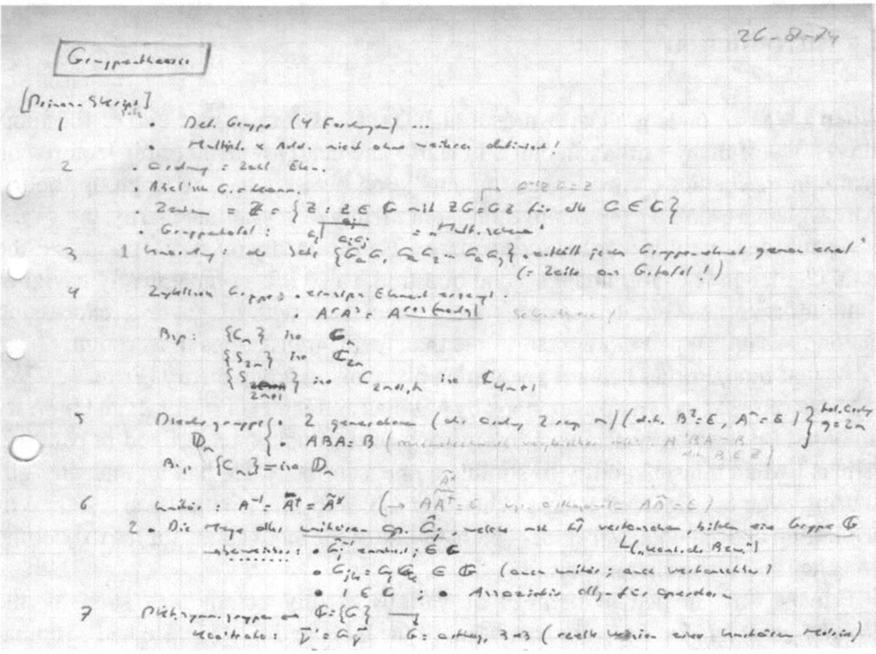

**Fig. 1** The beginning of eight pages of notes that I wrote in August 1974 while preparing my diploma examinations, based on Primas' lectures on group theory. Not many teachers inspired me to take such careful notes. I fear that none of my own students ever have

## 3 Primas' Seemingly Innocuous Riddles

Primas had the precious gift to raise one's curiosity. In high-resolution nuclear magnetic resonance (NMR) spectra, one frequently observes "multiplets" that have a fine structure due to so-called "scalar couplings", which are displaced with respect to each other by so-called "chemical shifts", as in Fig. 2a. Primas once asked an apparently naive question: Could one design an experiment that would cause the multiplets to collapse so as to keep only information about the chemical shifts? This seemingly innocuous riddle has become known as "the challenge of homonuclear decoupling".

Early attempts to meet this challenge relied on broadband irradiation of nearly the entire spectrum, except for a small window where one expected to observe simplified signals. Even today, this *Gedankenexperiment* seems a tall order, and it has never been carried out in practice. A more realistic approach was proposed by Richard Ernst and his co-workers, who suggested that one could calculate a skew projection of a suitable two-dimensional spectrum. Unfortunately, this idea turned out to be of limited practical use. Yet another variant was developed quite recently in our laboratory (Carnevale et al. 2012). This "polychromatic" approach allows one to achieve a stepwise simplification by multiple irradiations at predetermined frequencies (Fig. 2).

This is, at best, a small step towards Primas' vision. Despite its obvious limitations, we were so enthralled with our own ideas that we promptly dedicated a paper (Carnevale et al. 2012) to Hans Primas (Fig. 3) The acknowledgement of this paper reads:

> This communication is dedicated to Hans Primas, Professor Emeritus, ETH Zürich. H. Primas was one of the Ph.D. supervisors of R.R. Ernst, and first asked the question if one could decouple all homonuclear interactions in a high-resolution spectrum, so as to observe only chemical shifts. He also taught basic quantum mechanics to one of the authors (G.B.), and often made the point that asking good questions is a far greater challenge for scientists than providing answers.

**Fig. 2** Stepwise simplification of multiplets in a proton NMR spectrum by multiple irradiations at predetermined frequencies (Carnevale et al. 2012)

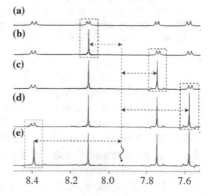

## Polychromatic Decoupling of a Manifold of Homonuclear Scalar Interactions in Solution-State NMR

Diego Carnevale,*[a] Takuya F. Segawa,[a] and Geoffrey Bodenhausen[a, b]

*Dedicated to Professor Hans Primas*

*12 August 2012*

*Lieber Herr Bodenhausen,*

*herzlichen Dank für Ihre mir gewidmete Arbeit, die ich mit Vergnügen gelesen habe.*

*Obwohl ich mich seit Jahren mit ganz anderen Dinge beschäftige, war es gut, wieder einmal an die schönen alten Zeiten erinnert zu werden. Auch erinnere ich mich, dass ich seinerzeit durchaus von Pulsmethoden Kenntnis hatte, aber die damit verknüpften experimentellen Schwierigkeiten als fast unüberwindbar betrachtete. Heute kann ich nur noch die hervorragenden Fähigkeiten (und Möglichkeiten) der Experimentatoren bewundern.*

*Viel Glück für Ihre weiteren Arbeiten!*

*Mit den besten Grüssen und nochmals vielem Dank*

*Ihr Hans Primas*

**Fig. 3** Title page of our paper dedicated to Hans Primas in 2012, and his reply

Primas was kind enough to send us a warm-hearted message by electronic mail (cf. Fig. 3):

> Dear Mr. Bodenhausen,
> Thanks for the paper dedicated to me, which I read with pleasure.
> Though I have been occupied with entirely different subjects for years now, it was nice to be reminded to the good old times. I remember that I was aware of pulse methods at the time, but considered the associated difficulties as almost insurmountable. Today I can only admire the outstanding accomplishments (and possibilities) of experimenters.
> Good luck for your future work!
> With best regards and thank you again,
> Yours, Hans Primas

Our recent attempts to separate ortho- and para-water (Mammoli et al. 2015), our work on triplet-singlet imbalance, on long-lived states, etc., may be considered as feeble attempts to breathe some life into the abstract world of irreducible representations. Figure 4 shows the nuclear spin states of a water ($H_2O$) molecule. In the days when I attended Primas' undergraduate lectures, I was puzzled that one could write sums and differences of spin states $\alpha$ and $\beta$ associated with nuclei that are not located on the same spot. Indeed, such spin states are not localized—this is one of the first mysteries that the young student must harness when entering the world of quantum mechanics. Primas used to recommend: simply start by working through the formalism, and "turn the handle" of the mathematical machinery, before asking

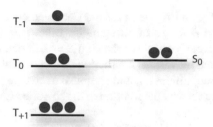

**Fig. 4** Symmetry-adapted nuclear spin states of a water ($H_2O$) molecule. The three triplet states $|T_{-1}\rangle = |\beta\beta\rangle$, $|T_0\rangle = N(|\alpha\beta\rangle + |\beta\alpha\rangle)$, and $|T_{+1}\rangle = |\alpha\alpha\rangle$ with the norm $N = 2^{-1/2}$ correspond to ortho-water, while the singlet state $|S_0\rangle = N(|\alpha\beta\rangle - |\beta\alpha\rangle)$ corresponds to para-water (Mammoli et al. 2015)

questions about "how" and "why". Today, I wonder how Primas would have talked about the distinction between *ortho-* and *para*-water. How would he have viewed our claims that we should be able to separate them physically?

## 4   Should Liouville Matrices be Called Primas Matrices?

So-called *superoperators* flow from a concept that, as far as I know, can be traced back to Primas, in particular to his famous paper with Banwell (Banwell and Primas 1963). Clearly, the concept was too difficult for undergraduate lectures, and I only started to grasp its implications around 1983 when working on *Principles*, as Richard Ernst's book with Alexander Wokaun and myself would come to be known. Superoperators act on operators, like operators act on quantum states. There is a hierarchy of spaces: while quantum states span a Hilbert space, operators span a Liouville space.

Many modern NMR methods could not have been designed without density operators. It has become standard practice to expand density operators in terms of a suitable basis set of operators (such as the ubiquitous Cartesian product operators, single-transition operators, irreducible tensor operators, etc.). Without the ability to construct Liouville matrices that map interconnections between these operators, the development of many modern NMR methods would not have been possible.

Let me give an example that is concerned with the 6-dimensional space spanned by the two levels of the spin $S = 1/2$ of a nitrogen-15 nucleus and the three levels of the spin $S = 1$ of a deuterium nucleus (Canet et al. 2016). One can conceive a set of $4 \times 9 = 36$ operator products but, for reasons of symmetry, only 9 are relevant to describe our experiments.

The following 9 products of angular momentum operators span a 9-dimensional Liouville space:

$$2\sqrt{2/3}N_y, \ 2N_xD_z, \ 2\sqrt{2}N_y(3D_z^2 - 2E), \ 2N_xD_y, \ 2\sqrt{2}N_y(D_zD_y + D_yD_z),$$
$$2\sqrt{2}N_y(D_x^2 - D_y^2), \ 2N_xD_x, \ 2\sqrt{2}N_y(D_zD_x + D_xD_z), \ 2\sqrt{2}N_y(D_xD_y + D_yD_x).$$

The symbols $N_x$, $N_y$, and $N_z$ stand for the Cartesian components of the angular momentum of the spin $S = 1/2$ of nitrogen-15. Likewise, $D_x$, $D_y$, and $D_z$ stand for angular momentum components of the spin $S = 1$ of deuterium. The 9 operator terms are interconnected by various interactions, such as scalar $J$ couplings, radio-frequency fields with amplitudes $\omega_1$, and chemical exchange processes with rates $k$. Their effect can be described by the solution of the so-called Liouville–von Neumann equation,

$$\sigma(t) = \exp(-Lt)\,\sigma(t = 0),$$

where $\sigma(t)$ is the density operator and $L$ the Liouville superoperator, which can be represented by a $9 \times 9$ dimensional matrix in the basis of the 9-dimensional Liouville space mentioned above:

$$\begin{pmatrix}
0 & 2\sqrt{\tfrac{2}{3}}J\pi & 0 & 0 & 0 & 0 & 0 & 0 & 0 \\
-2\sqrt{\tfrac{2}{3}}J\pi & k & -\tfrac{2}{\sqrt{3}}J\pi & \omega_1^D & 0 & 0 & 0 & 0 & 0 \\
0 & \tfrac{2}{\sqrt{3}}J\pi & k & 0 & \sqrt{3}\omega_1^D & 0 & 0 & 0 & 0 \\
0 & -\omega_1^D & 0 & k & -J\pi & 0 & \Omega_D & 0 & 0 \\
0 & 0 & -\sqrt{3}\omega_1^D & J\pi & k & -\omega_1^D & 0 & \Omega_D & 0 \\
0 & 0 & 0 & 0 & \omega_1^D & k & 0 & 0 & 2\Omega_D \\
0 & 0 & 0 & -\Omega_D & 0 & 0 & k & -J\pi & 0 \\
0 & 0 & 0 & 0 & -\Omega_D & 0 & J\pi & k & -\omega_1^D \\
0 & 0 & 0 & 0 & 0 & -2\Omega_D & 0 & \omega_1^D & k
\end{pmatrix}$$

Thus, Liouville superoperators can be represented by matrices that have the dimensions of the corresponding Liouville space. Such matrices have become a mainstay of modern magnetic resonance. It is not quite clear why matrices of this type have been attributed to the French mathematician Joseph Liouville (1809–1882). I like to think that it would be more appropriate to speak of "Primas matrices". One wonders who detains the authority to brand such names.

The formalism built on these principles allowed us to design new experiments that make it possible to determine the exchange rates of deuterium ions $D^+$ that hop between secondary amines ($R_1R_2ND$) and heavy water molecules ($D_2O$), and to compare them with analogous exchange rates involving protons $H^+$:

$$\text{N-D} + \text{D}'^+ \longrightarrow \text{N-D}' + \text{D}^+ \quad \text{rate } k_D$$
$$\text{N-H} + \text{H}'^+ \longrightarrow \text{N-H}' + \text{H}^+ \quad \text{rate } k_H$$

Not surprisingly the exchange rates $k_H$ and $k_D$ (which vary from $20\,\text{s}^{-1}$ to $20000\,\text{s}^{-1}$ depending on temperature and acidity) are affected by kinetic isotope effects. Vladimir Prelog would no doubt have liked to discuss such effects with Hans Primas in his early days!

I remember that Jean Jeener, one of the founding fathers of modern magnetic resonance, once gave a talk about these matters. At that time (about 1990) Jeener was of the opinion that, while density matrices undeniably exist, there is no such thing as a density operator. For Primas, a density matrix is merely a matrix representation of an abstract object called density operator. It would have been wonderful to attend a debate between Jeener and Primas on these issues!

## 5 Images as a Stimulus of Imagination

Primas had decided to leave the field of magnetic resonance long before imaging (MRI) was invented in 1973. If we could have asked him, Primas would probably have bluntly replied that he had not contributed in any way to this spectacular development. I would be tempted to challenge that view. Primas' ideas about instrumentation and the role of Fourier transformations (Primas and Günthard 1958) pervaded the field. Figure 5 reveals the "plumbing" of the arteries in my brain, presented in such a manner that the image appears brighter when the blood flows faster. Note that there is hardly any visible vascularization in the cortex, because capillary flow is too slow to be observed.

**Fig. 5** Angiograph of the author's brain (recorded in April 2014 at the hospital of La Salpêtrière, Paris)

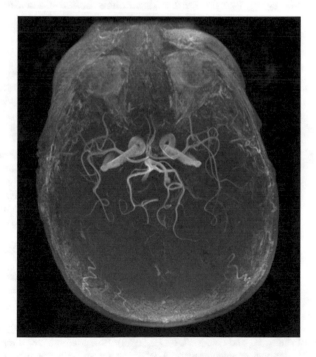

This leaves plenty of scope for imagination. Other MRI pictures can show contrast due to local variations in relaxation rates, anisotropic diffusion tensors, magnetic susceptibility, proton exchange rates between water and tissue, and local variations of the rates of interconversion between oxy- and deoxy-haemoglobin, as occur in "functional" MRI. No doubt C.G. Jung (1944), whose writings were a source of inspiration for Primas for decades, would have been inspired to construct exciting hypotheses about such pictures. We can speculate that Primas too would have liked to discuss the promises and limitations of MRI.

The symposium in honor of Hans Primas that was held at the Collegium Helveticum on November 27th, 2015, was built on the idea that one could trace his intellectual "legacy". This concept has many fascinating and controversial aspects. Fascinating because one would like to understand how novel ideas arise, how they can be passed on from generation to generation, how they often wither and fade away, but on rare occasions blossom and propagate.

Yet the concept of an intellectual legacy is controversial since legacies can follow tortuous ways. An idea rarely appears *ex nihilo*, for its inventor is deeply immersed in his culture. He may have had teachers, peers, and, perhaps most importantly, students. One may attempt to trace legacies using the supposedly objective tools of "bibliometrics", on the assumption that, if a master has inspired a student, the latter would cite the former in his own writings. For my part, I fear that I have hardly ever cited any of Primas' papers in some four decades of research and publishing, so there is little material evidence of any filiation. Yet I gratefully acknowledge *my debt to his legacy, was ich seinem Vermächtnis schulde, ma dette à son héritage.*

# References

Banwell, C.N., and Primas, H. (1963): On the analysis of high-resolution nuclear magnetic resonance spectra. I. Methods of calculating NMR spectra. *Molecular Physics* **6**, 225–256.

Canet, E., Mammoli, D., Kadeřávek, P., Pelupessy, P., and Bodenhausen, G. (2016): Kinetic isotope effects for fast deuterium and proton exchange rates. *Physical Chemistry Chemical Physics* **18**, 10144–10151.

Carnevale, D., Segawa, T.F., and Bodenhausen, G. (2012): Polychromatic decoupling of a manifold of homonuclear scalar interactions in solution-state NMR. *Chemistry. A European Journal* **18**, 11573–11576.

Mammoli, D., Salvi, N., Milani, J., Buratto, R., Bornet, A., Sehgal, A.A., Canet, E., Pelupessy, P., Carnevale, C., Jannin, S., and Bodenhausen, G., (2015): Challenges of preparing, preserving and detecting para-water in bulk: Overcoming proton exchange and other hurdles. *Physical Chemistry Chemical Physics* **17**, 26819–26819.

Mathiak, K., and Stingl, P. (1968): *Gruppentheorie für Chemiker, Physiko-Chemiker, Mineralogen*, Vieweg, Braunschweig.

Müller, N., Bodenhausen, G., and Ernst, R.R. (1987): Relaxation-induced violations of coherence transfer selection rules in nuclear magnetic resonance. *Journal of Magnetic Resonance* **75**, 297–334.

Jung, C.G. (1944): *Psychologie und Alchemie*, Rascher, Zürich.

Primas, H. (1978): *Elemente der Gruppentheorie*, Verlag der Fachvereine, Zürich.

Primas, H., and Günthard, Hs.H. (1958): Eine Methode zur direkten Berechnung des Spektrums der von quantenmechanischen Systemen absorbierten bzw. emittierten elektromagnetischen Strahlung. *Helvetica Physica Acta* **31**, 413–434.

# Theoretical Chemistry and More: Personal Annotations to Hans Primas and His Work

Ulrich Müller-Herold

**Abstract** In the mid 1960s, Hans Primas concentrated on research into the enigmatic relation between chemistry and quantum mechanics: How can a molecule exhibit purely classical features as in stereochemical ball-and-stick models alongside purely quantal properties as in chemical spectroscopy? Due to the discovery of superselection rules in the 1950s Primas was able to propose a solution in terms of classical observables. In this vein he contributed to the theory of chirality and to the measurement problem of quantum mechanics. In addition, he initiated research on elementary systems and the construction of observables in general. At the end of the 1970s, a permanent discussion topic in the Primas group was reductionism: How can a given theoretical description be related to more fundamental lower-level theories? Primas' magnum opus *Chemistry, Quantum Mechanics and Reductionism* of 1981 addresses this question, which is difficult and controversial at the same time. After his retirement, Primas restarted earlier work on time and irreversibility. This culminated in a seminal paper on "Time-Entanglement between Mind and Matter" in 2003 that explores Wolfgang Pauli's idea that mind and matter are complementary aspects of the same reality.

## 1 Introduction

When Hans Primas graduated from the chemical branch of the Winterthur Technical School, the practical part of his final examination was the synthesis of chloramphenicol, an antibiotic discovered in 1947. The organic chemistry community he thereby entered was a largely self-sufficient universe, historically shaped by great achievements in structure determination and synthesis of natural compounds. It was a world of autochthonous classical concepts with dash formulae and ball-and-stick models as the main paper tools (Klein 2003), with only poor links to physics. "The only thing I need from physics is a laboratory balance" was a popular saying by ETH Zurich's Nobel Laureate Leopold Ruzicka (1887–1976).

U. Müller-Herold (✉)
Department of Environmental Systems Science, ETH Zurich, Zurich, Switzerland
e-mail: mueller-herold@env.ethz.ch

© Springer International Publishing Switzerland 2016
H. Atmanspacher and U. Müller-Herold (eds.), *From Chemistry to Consciousness*, DOI 10.1007/978-3-319-43573-2_3

23

After World War II the situation changed dramatically: In the wake of military radar technology new spectroscopic methods arose, based on quantum mechanics and progressively invading everyday chemistry. From the early 1950s onwards, chemists found themselves in situations in which quantum-mechanical methods could be of use to them, methods that cannot be understood in terms of classical concepts.[1]

This was the situation Primas encountered when he arrived to work at ETH's Laboratory for Organic Chemistry in 1953, one of the world's leading centers for structural chemistry, where he joined the group of Hans Heinrich Günthard who introduced post-war chemical spectroscopy at the ETH. Professor Günthard's aim was to increase the sensitivity and resolution of nuclear magnetic resonance (NMR) discovered by Purcell and Bloch in 1946, so that NMR spectroscopy could be used for chemical structure determination.

Largely on a self-taught basis, Primas trained himself to be a professional electronics engineer and rapidly achieved great success in spectroscopic engineering. Under the aegis of Professor Günthard he built the first high-resolution NMR instruments for chemical analytics, registered 30 patents, and solved numerous hard- and software problems of early NMR: "In ten years, he achieved more than other successful scientists create during a lifetime" (Ernst 1999, p. 35).

## 2   Preludes in Theoretical Chemistry

From the very beginning, the interconnections between chemistry, biology and physics held far more fascination for Primas than potentially extending the fields of application for NMR. Even as a teenager, he had been inspired by Bernhard Bavink's *Results and Problems in the Natural Sciences. An Introduction to Current Natural Philosophy*, a German book that was published in nine updated editions between 1914 and 1949 and reported on the current state of the field. For Primas, the natural sciences were a unified whole, which should not be divided into subdisciplines nor into an empirical and a theoretical branch. Thus, it was only a matter of time before he abandoned the familiar context of spectroscopic engineering to address more fundamental questions of natural science.

When Primas entered the scene, theoretical chemistry had just had its first encounter with electronic computers. As early as 1927, Heitler and London had given a qualitative explanation of the chemical bond in the hydrogen molecule $H_2$ with the help of the new quantum mechanics. Thereby they solved a problem that had not been solvable within the old Bohr–Sommerfeld quantum theory (Heitler and London 1927). However, it was not until 1960 that Kolos and Roothaan published the first quantum-chemical calculation on electronic computers (Kolos and Roothaan 1960). They used two main technical tools: the Born–Oppenheimer or

---

[1]Roughly speaking, an observable is called classical if there is no other observable such that their joint dispersion-free measurement is impossible. Otherwise it is quantum mechanical ("quantal").

clamped-nuclei approximation, which in a first step led to a Schrödinger equation for the electrons alone, and the molecular orbital concept to solve this equation.

In his 1964 inaugural publication "What Are Electrons?" Primas (1964) addressed the problem that chemists use molecular orbitals in an intuitive creative way that is apparently not covered by quantum mechanics. Chemical electrons are individual, i.e., they can be distinguished and numbered: A chemist speaking of the carbon 1 s electron thereby means an electron described by a distinct one-particle state function, a so-called 1 s orbital. In physics, on the other hand, electrons are indistinguishable. They can be counted yet not numbered, due to Pauli's antisymmetry requirement for the wave function.

Primas then presented a surprising solution of the disagreement: He pointed out that the poles of the so-called single-particle Green's function (Layzer 1963; Migdal and Larkin 1964) essentially correspond to what he called the chemists' quasi-electrons.[2] The $N$ quasi-electrons of a molecule are described by the first $N$ so-called natural orbitals. The chemists' concept of electrons thus can be rigorously justified: *If they say electrons they mean orbitals.* Quoting Michael Polyani ("there are things that we know but cannot tell"), Primas formulated an instructive message that went right to the chemists' heart. In this inaugural paper he spoke more simply and directly to the chemistry community than he ever would again.

In 1967, he took a radical decision: As a full professor of physical and theoretical chemistry at the ETH, he formally left the field of NMR to Richard Ernst, in order to concentrate on research into the relation between chemistry and quantum mechanics: How can a molecule exhibit purely classical features in stereochemical ball-and-stick models alongside purely quantal properties as in chemical spectroscopy? In his Elmau talk, Primas (1968) sketched his research program for the years to come: revision of the foundations of quantum chemistry. The talk begins with a list of several problems, which cannot be solved by a Schrödinger equation. It then refers to a new algebraic formulation of quantum mechanics, which had been developed by Haag and Kastler (1964) and others in connection with quantum field theory and provided mathematically rigorous solutions to the above problems. In the last section of his paper, Primas comes to the technical core of his project: to develop extended quantum chemistry in the language of algebraic quantum mechanics. On account of the fundamental nature of the project it was clear from the beginning that this had to include an analysis of the mathematical formalism, its various interpretations and the measurement problem.

---

[2]The electronic single-particle Green's function of a molecular system can be directly calculated from an integro-differential equation, which avoids having to solve the electronic Schrödinger equation.

## 3   Classical Observables in Molecular Quantum Mechanics

Chemical systems are partly quantal and partly classical, so they do not share the simplicity of purely quantal and purely classical systems. In one and the same object, quantal and classical features coexist and interact with each other. In the early 1960s Primas had realized that naive applications of traditional quantum mechanics to such systems give no reasonable description, and that von Neumann's irreduciblity postulate had to be replaced in one way or another.[3]

With respect to von Neumann's irreducibility postulate, Primas made an important step further through a fortunate coincidence. In the 1960s his research was partly funded by grants of the Swiss National Science Foundation. The reviewer in charge of his proposals was Josef Maria Jauch, a prominent theoretical physicist from the University of Geneva. In 1961 Jauch and his coauthor Badyanath Misra had shown (Jauch and Misra 1961) that in systems with superselection rules[4] a special class of operators occurs, which they called essential observables, commuting with all other observables. They describe the classical part of a system. In May 1970, during a long walk on the Zürichberg hill Jauch convinced Primas that the solution to his problem of a unified simultaneous treatment of classical and quantal properties might lie in the concept of superselection rules and essential observables. Primas later named these *classical* observables and they were the clue to what he had sought for so long.

He immediately realized that classical observables could bridge the socio-cultural gap between chemistry and quantum mechanics. Chemists are not happy with the traditional statistical and epistemic interpretation of quantum mechanics since it disagrees with their view on the nature of material objects. By tradition, chemical theories have an ontic interpretation.[5] That is, the referents of their theories are single systems together with their objective properties. In their view, anything which is practically real should at least possibly be taken as objectively real.

For a chemist, there is no difference in principle between a tartaric acid crystal and a single molecule of the amino-acid alanine, which also exists in a left- and a right-handed form. If the chirality of tartaric acid crystals is accepted as a real objective property, independent of our knowledge or measurement, then one should also accept the chirality of a single molecule of alanine as a real objective property.

---

[3]The postulate implies that for each observable there is a least one other non-commuting observable making their joint dispersion-free measurement impossible. This excludes the existence of classical observables.

[4]For details on superselection rules see the contribution by Domenico Giulini in this volume.

[5]The interpretation of a theory is *epistemic* if the theory is regarded as dealing with the *knowledge* of the experimenter on systems he deals with; it is *ontic* if it tells how the world *is*. The proper place for epistemic theories is engineering, the objective of ontic theories is to understand the fundamental nature of things.

# 4 The Riddle of Molecular Structure and Chirality

The concept of chemical structure is a classical idea that is foreign to traditional quantum mechanics. When Primas took it up, it already had a long pre-history and, hence, a variety of different meanings. For clarity, I will refer to what an organic chemist has in mind when speaking about the structure of molecules such as benzene or methane. In a consistent theoretical description, it comes into being through the lowering of some symmetries. Usually this is smuggled into quantum chemistry via the Born–Oppenheimer approximation to the molecular Schrödinger equation.

Although the Born–Oppenheimer approximation leads to correct and physically meaningful results—and although it can be mathematically justified—on a more fundamental level the quasi-classical nature of nuclear position is not explained in this way. From the algebraic perspective, the situation remains unchanged: we see a lowering of symmetry but do not know in physical terms where it comes from. This central question remains unanswered.

When Primas realized that the riddle of molecular structure might be a nut too tough to crack, he heeded the advice of the distinguished Hungarian mathematician George Polya: If there is a problem one cannot solve, there is often a simpler one that one cannot solve either. Polya then suggests: Find it! The basic idea is obvious: With a problem that is simpler but structurally similar, one may perhaps move forward a little and find some sort of clue for the treatment of the original problem. The apparently simpler problem is the following: How can the existence of optical isomers—in the chemist's language, enantiomers or chiral molecules—be reconciled with the first principles of quantum mechanics?

In crystallized form or dissolved in optically non-active solvents, chiral substances rotate the plane of polarization of plane-polarized light. The standard example used by Primas and his followers is the simplest chiral amino-acid alanine: $CH_3CH(NH_2)COOH$. For this molecule, the Hilbert-space model of traditional quantum chemistry predicts a space-reflection invariant, i.e., optically inactive, non-chiral ground state. This prediction contradicts all experimental findings: Space-reflection invariant states of alanine do not exist. The experimentally observed states of lowest energy arise in pairs and are chiral, i.e., the molecules exist only in left- or right-handed forms $\Psi_+$ and $\Psi_-$ that are transformed into one another by a space-reflection $U$ (see Fig. 1).

Primas conjectured that the disappearing of space-inversion symmetry is due to a coupling of the molecule to the transverse part of the quantized electromagnetic radiation field. He made a minimal *ansatz* and hoped that the resulting one-spin-many-bosons type model would be simple enough for closed analytical solutions and a detailed study of the symmetry lowering in the limiting case of infinitely many field bosons.

In 1974 his Ph.D. student Peter Pfeifer, a chemist with a strong background in theoretical physics and the talents of a professional mathematician, started to work on the model. For six years, the model resisted the common efforts of Pfeifer and Primas to find an exact solution. However, turning towards approximate solutions on

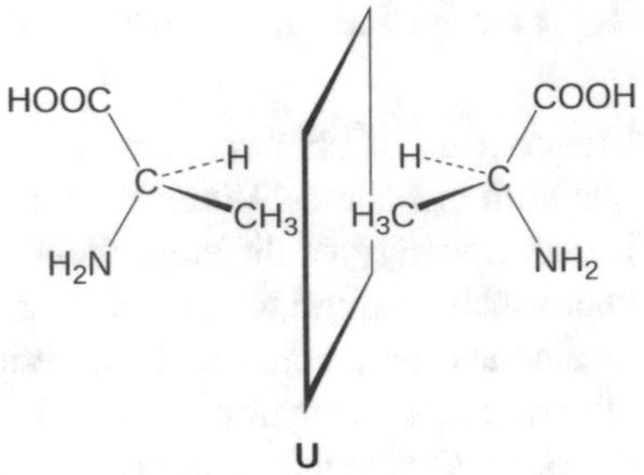

**Fig. 1** The *left-* and
*right*-handed forms of
alanine are transformed into
one another by a
space-reflection $U$:
$$U\psi_+ = \psi_-, U\psi_- = \psi_+$$

the basis of the Hartree factorization, Pfeifer succeeded in deriving groundbreaking results: For very small energy differences between the energetically lowest lying two states of the isolated molecule (which means that the field is switched off) the ground state splits under the influence of the radiation field (if the field is switched on again) into two chiral molecules that are mirror images of each other. They differ through handedness, a new classical observable.

When Pfeifer (1980) presented his thesis, it opened a novel route to a long-standing problem and brought him the appreciation of prominent authorities in mathematical physics (Wightman and Glance 1989). In the long and medium term, however, Pfeifer's results turned out to be controversial. Alongside confirmation there was also criticism, amongst others with the argument that the Hartree approximation as a rule gives phase transitions too easily. Within the Primas group, Pfeifer's work was taken up by Anton Amann who translated it into the formalism of algebraic quantum mechanics. He arrived at significant improvements without ultimately solving the "paradox of optical isomers".

# 5   Joining Primas' Program

As a medical doctor with a chemistry diploma I started working as Primas' assistant in 1973. He welcomed me with a half-semester course on errors in books on quantum mechanics. Before entering the sphere of advanced quantum theory I needed a warm-up. To start with, Primas gave me a seemingly innocent mathematical problem: What are the *mathematical* reasons that the differential equations of chemical kinetics never lead to negative concentrations? Fortunately, I managed to answer the question within an acceptable period of time (Müller-Herold 1975).

In 1977 he invited me to co-author the study *Quantum Mechanical System Theory* (Primas and Müller-Herold 1978), a hundred-page paper on observations and stochastic processes in quantum mechanics, for which he wrote the general theory part and I was in charge of the examples. This was my initiation into algebraic quantum mechanics. Also published in 1977 was the seminal paper on the chemical potential in algebraic quantum mechanics by Araki et al. (1977), four of the leading figures in the field at the time. For me, this was a paper of unprecedented mathematical complexity. Its central ingredient, the so-called non-commutative cocycle Radon–Nikodym derivative, had been introduced four years earlier by Fields Medal winner Alain Connes in his 1973 dissertation. Since it seemed to open a route to novel applications of algebraic quantum mechanics in chemistry, I saw a chance to find my own way in the field.

As a first exercise I examined a problem that Araki et al. (1977) had not written about, probably because the answer was too clear to them: Is the chemical potential they had derived a classical observable? In 1980 I published a paper with an affirmative result (Müller-Herold 1980). Around that time, I was proposed for membership in the International Association for Mathematical Physics. Thus encouraged, I started a more ambitious project on the algebraic theory of the chemical potential in the presence of chemical reactions. But how should one implement chemical reactions in the framework of algebraic quantum mechanics? The answer had to be formulated in the language of automorphisms of algebras of observables. After some technical excursions into crossed products of W*-algebras with compact abelian groups and the Galois theory associated with them, this battle of strength ended with a central formula of traditional chemical thermodynamics: the well-known condition of reactive equilibrium (Müller-Herold 1982, 1984).

It was a heroic achievement and it won me the admiration of Primas' master pupil Anton Amann: Isn't it crazy that from the stratospheric regions of a most advanced fundamental theory mundane equations of physical chemistry can be recovered? Otherwise, however, the response was disappointing. Although I was awarded a prize when the paper was accepted as a *habilitation* thesis by the chemistry department of the ETH, the distinguished chemist Duillio Arigoni asked: "What should he be giving lectures about, when even we cannot understand him?" Still more disappointingly, the reaction from the physics community was lukewarm. No less a figure than Huzihiro Araki (1983, 1986) completely missed the chemical point in his reports on the paper for *Mathematical Reviews*.

Altogether I had to cope with the following situation: The paper was interdisciplinary to a degree that chemists at best understood the problem, but even the upper circles of theoretical chemistry were ignorant of the formalism and reluctant to take note of its alphabet. With the mathematical physicists, meanwhile, it was the other way around. In 1983 there was no scientific community willing to discuss chemical problems with algebraic quantum mechanics; the gulf in between was simply too large. I never raised this admittedly delicate point with Primas. Might it not be that a similar analysis applied to the whole of his research program?

For me at least, it was evident that on this basis it would not be possible to find a suitable academic position in theoretical chemistry. The same became true for all of Primas' Ph.D. students: Those who stayed in academia became professors of mathematics, experimental physics, mathematical physics, and medical research, however none of them have done so in chemistry. Consequently, I redirected my professional activities towards environmental sciences. From fascinating things I moved to urgent ones.

# 6 Measurements

The so-called measurement problem in quantum mechanics was not solved but only clearly posed by John von Neumann (1932), who formalized the unitary time evolution given by the Schrödinger equation but also proposed a second type of state change, which he claimed to occur during a measurement. Formally, this *reduction of the wave packet* rests upon the projection postulate that von Neumann added to the formalism of quantum mechanics. Within the traditional formalism, the reduction of the wave packet cannot be described by a Schrödinger equation. Historically, this gave rise to raging philosophical controversies regarding the interpretation of quantum mechanics, including its paradoxes. These crucially depend on the fiction of *measurements of the first kind*, which are defined as instantaneous, repeatable and giving sharp values.

As Primas had always highlighted, all measuring processes one *really* understands are not measurements of this type, and they can be discussed without the projection postulate. In contrast to fundamental quantum mechanics, quantum electronics has a well-developed measurement theory, created in the context of detection and estimation problems in radar systems and optical communication with lasers. In engineering quantum electronics, the simultaneous measurement of non-commuting observables is everyday practice.[6] In science, this led to a dichotomy: The theoretical physics and philosophy of science communities worked on the "measurement problem of quantum mechanics", and experimenters and engineers on the technology of "real" experiments. Primas formulated and partially also published contributions to both.

Throughout his professional life, he struggled with the measurement problem of quantum mechanics, on one hand, because—as in chemistry—it involved an interplay of classical and quantal elements and, on the other, because it ignited a controversy about the interpretation of quantum mechanics, which ranks as one of the great philosophical debates in the history of science. What was new here was particularly the observation that there are various interpretations of the formalism of quantum mechanics, which itself is broadly undisputed, that are consistent in themselves but nonetheless incompatible with one another: the Copenhagen interpretation, the von Neumann-London-Bauer interpretation and the Everett interpretation, to which Primas dedicated an informative overview in his monograph of 1981.

At the technical level, Primas contributed to the measurement problem with an unpublished model (Primas 1969), which survived only as a citation in a seminal work by Hepp (1972, p. 246). When the measurement problem (in the setting of Hepp) was finally solved by Lockhart and Misra (1986), Primas (1987) expressed his appreciation in *Mathematical Review*:

> This paper is a breakthrough, although it does not yet solve all problems related to measurement processes in the statistical interpretation of quantum mechanics. But it shows conclusively that the so-called measurement problem of quantum mechanics is neither a

---

[6]Optical heterodyning, for example, is equivalent to an optimal measurement of the Schrödinger pair $(P, Q)$ or, equivalently, of the photon annihilation operator $Q + iP$.

pseudoproblem nor a philosophical question, but a well-posed problem of mathematical physics which can be solved in the framework of algebraic quantum mechanics.

As a designer of high-resolution NMR instruments, on the other hand, Primas was familiar with engineering-type real measurements. He considered NMR as a paradigm for quantum experiments in general, and Bloch's equations of 1946 as the historically first example of a positive dynamical semigroup—a structure that would receive recognition 30 years later as fundamental for the description of experiments involving open quantum systems. He spent a long time attempting to generalize Bloch's equations to a theory of positive semigroups. This was first achieved by Lindblad (1976) and Gorini et al. (1976) on the basis of Stinespring's (1955) concept of "complete positivity". Together with his Ph.D. student Guido Raggio, Primas later showed that Bloch's equations are actually a completely positive semigroup in the sense of Lindblad and Gorini et al. (Raggio and Primas 1984).

The description of experiments with the help of completely positive semigroups led to the theoretically fundamental *response theory*, which replaced Fermi's historically significant yet merely heuristic and theoretically refutable "golden rule". Primas stressed the view that all quantitative experiments can be described using the modern response theory of open systems. Bloch's description of NMR is merely an especially simple case of this theory.

# 7 Writing Books

At the end of the 1960s, the US publishing company Academic Press approached Primas with an offer to publish his critical revision of the foundations of quantum mechanics in book form. The project ended without success, because Primas spent a longer time working on it than had been agreed with the publishers. A second, and also unsuccessful, attempt at publication with Germany's chemical flagship journal *Angewandte Chemie* did, however, result in an offer from Prof. Werner Kutzelnigg, one of the editors of *Lecture Notes in Chemistry* at Springer, to publish the material there. With a foreword by the philosopher of science Paul Feyerabend it appeared in 1981 under the title *Chemistry, Quantum Mechanics and Reductionism* (Primas 1981) as a monograph of 451 p.

"The purpose of this book", Primas wrote in the introduction, "is to provide a deeper insight into modern theories of molecular matter. It … is not meant to be a textbook … in many respects it has complementary goals". The book commences with a first chapter on open problems in present-day theoretical chemistry. In the second chapter, Primas gives a general account of the structure of scientific theories. Couched in Jungian language, it presents an overview of the sources, achievements and limitations of present-day scientific theories. The third chapter reviews the status of pre-war quantum mechanics, including its alternative mutually incompatible statistical interpretations, which do not violate the experimental facts. It becomes clear that Primas is dissatisfied with the different versions of the Copenhagen interpretation and prefers that of von Neumann, London and Bauer, from which he merely wants to remove the outdated irreducibility axiom excluding classical properties.

In Chaps. 4 and 5, he develops the modern codifications of quantum mechanics: the algebraic approach, the quantum-logic approach and the convex-state approach. He then designs a framework for theoretical chemistry, a merger of quantum logics and algebraic quantum mechanics, which forms the core of the book. Within this framework, he touches on various open questions: theory reduction, holism, complementarity, the ontic interpretation of quantum mechanics, and the existence of isolated quantum systems.

The book is difficult and provocative at the same time. Between enthusiasm and hostility there were reactions of every shade. But, as Paul Feyerabend (1983) put it, "it is interesting to see how some reviewers work their way through the initial shock to reach a positive judgement. But there is no escape from your arguments". In particular, for Anglosaxon reviewers, the Kantian spirit of ontic interpretations and of a Jungian world view was unaccustomed, to say the least. One of them, however, wrote at the end of a long and well-balanced review (Sutcliffe 1983):

> It should not be forgotten that the ideas of Hamilton, which were contributory to Schrödinger's development of quantum mechanics, arose from Hamilton's philosophical position. Hamilton was a Kantian. (His) formulation of the optical-mechanical analogy was a consequence of his philosophy not any experiments.

I think that this reviewer has captured the gist of a scientific legacy: the Hamilton function slept for almost one hundred years. It came back as the Hamilton operator and turned out to be a key element in Schrödinger's discovery.

In his review, one of the true grandmasters wrote (Longuet-Higgins 1981):

> In this long and fascinating book the author reveals a deep understanding of the most difficult problems of quantum mechanics and its interpretation …He reveals himself as a master not only of the mathematics but also of the philosophy of science, and as a man who cares passionately about intellectual integrity, in the widest possible sense of the word… The book may be difficult, but it is quite unusually honest and thoughtful, and is likely to prove an invaluable antidote to the narrowness of outlook which is so often apparent in chemical theorizing.

And the inorganic chemist C.K. Jörgensen (1982) from Geneva concluded his review with a really astounding statement: "Nobody who has been reading this book can ever sleep entirely quietly".

Shortly later the mathematician Eduard Stiefel[7] came across Primas in the underground garage at the ETH. (Before the establishment of the *Dozentenfoyer* in the 1980s, this was the only place at the ETH where professors from different departments met informally on a semiregular basis.) Stiefel casually explored the possibility of an introduction to quantum chemistry at the undergraduate level for the *Teubner* publishing company in Stuttgart. In reaction to his suggestions, Primas asked me to coauthor such a book, on the grounds that it might be useful for my academic advancement.

---

[7] Known for Stiefel-Whitney cohomology, work on Lie groups and the early establishment of numerical mathematics and computer sciences at the ETH in 1949.

We rapidly agreed upon the aim: an elementary book, without recourse to higher mathematical tools such as group theory or functional analysis, which avoids the errors and detours of the historical development: no particle-wave dualism, no double-slit experiment, state functions instead of wave functions, etc. Needless to say, even at this elementary level the correct mathematical terminology had to be used—self-adjoint operators instead of Hermitian nes, density operators instead of density matrices—and didactic lies were strictly excluded: spin as a result of rotating electrons, etc. Conceptually, on the other hand, it was an attempt to bring even beginners closer to the connections with mathematics, philosophy and culture as a whole.

The resulting book *Elementary Quantum Chemistry* (Primas and Müller-Herold 1984) starts with a statement the first part of which came from Primas, and the second part from the coauthor: "Quantum theory is one of the great cultural achievements of our century and a part of general education for all those who possess the mathematical knowledge to understand it". It ends with an outlook on problems, which on the basis of the first principles of quantum mechanics had not been solved so far, among them the question that has been concerning me for the last fifteen years (Müller-Herold 2015): What precisely is the molecular structure of liquid water and how is it related to individual $H_2O$ molecules?

The book was a reasonable commercial success. In 1988, *Teubner* published a second edition and the book is still the subject of some second-hand trade. Recently, it was selected for the Springer Book Archives and will thus remain available on demand (Fig. 2).

**Fig. 2** From *right* to *left* Primas, the author and the inorganic chemist Professor Walter Schneider at Cortona (1987)

# 8    Coffee Conversations

Primas used to take afternoon coffee with his research group. Usually, he would bring the newly arrived preprints along, which always ignited discussion. Beside scientific issues, there were also discussions that drew more upon the current concerns of the times. Thus, it was that the word "motivation" first appeared in German-language conversations during the 1970s, and it became a habit in academia to approach any and everyone on the subject of what one's actual motivation as a scientist might be. Primas' response to this was indignant: "On the one hand, it is like solving crossword puzzles, and on the other, I shall not tell you". Later, however, he did begin slowly and cautiously to address the question. He mentioned the Indian mathematician Sriniwasa Ramanujan whose theorems were inspired by the Hindu goddess Yamagiri, the composer Johannes Brahms who received the themes of his compositions whilst in a trance-like state, and the mathematician Henri Poincaré who dreamt the solution to the problem of automorphic functions in a symbolic encoding.

One of the Ph.D. students said that scientists want to be famous and be awarded the Nobel Prize, to which Primas replied:

> Sometimes that might be the case, but it is of no particular interest and it gives a false impression of what we are doing. When Kepler discovered that the radii of planetary orbits fit only approximately into the Platonic bodies and embarked thereafter upon a 15-year search for a new theory, there was no fame and no Nobel Prize to be won.

Eventually he added (cf. Müller-Herold 2005, p. 56):

> All new ideas enter our awareness like a lightning strike. We do not know where these ideas come from and therefore say "from the unconscious". In earlier times, it would have been natural to refer to God as the source. The unconscious contains many things, including the remains of days, and that which is subliminally perceived, forgotten and repressed. This personal aspect of the unconscious is however not primarily decisive for the great ideas of science and the great inspirations in art. The great inspirations are unmediated expressions of the non-personal collective unconscious. The existence of a non-personal source for great inspirations is guaranteed by many thousands of years of experience, so that we may regard the reality of the collective unconscious as an empirical fact. The fascination that creative research holds for many natural scientists has its origins precisely in their personal participation in the wealth of the collective source.

With arguments such as these, he entered the territory of C.G. Jung's analytical psychology. Indeed, much earlier, in the vibrant atmosphere of early post-war Zurich, Primas had come across the ideas of Jung. Alongside his study of works such as *Psychology and Alchemy*, he also made direct contact with the Jung Institute. A connection of a very different kind came about because Primas and Jung had the same family doctor in Küsnacht near Zurich, where both of them were living, meaning he also obtained informal and anecdotal knowledge of Jungian circles.

At the end of the 1970s, C.G. Jung's encounters with Wolfgang Pauli became a new and ever more important subject of discussion: That the memorable dreams at the beginning of Jung's *Psychology and Alchemy* actually came from Pauli in the course of his analysis with Jung's student Erna Rosenbaum, and that Jung and Pauli

(1952) wrote a book together and maintained a regular exchange of letters for twenty-five years. Yet their correspondence remained largely unpublished in the archives of CERN in Geneva and at the ETH in Zurich, and the thoughts contained therein might become important for the future development of the natural sciences.

Another permanent item was reductionism: How can a given theoretical description be derived from more fundamental lower-level theories? Evidently, the relation between chemistry and quantum mechanics is a special case of the problem. Philosophical logics of science had proposed the Hempel–Oppenheim scheme as a solution to the general case. Primas was sceptical about the Hempel–Oppenheim proposal. After more than a decade of careful analysis, he arrived at the result that it does not work except in trivial cases (Primas 1991).

By the mid 1980s, environmental problems emerged as a new and important theme. At the ETH in Zurich, Primas worked in the background, albeit to great effect, with the establishment of a diploma curriculum in environmental sciences. He had his doubts as to whether a branch of science that should be held partly responsible for the degradation of the environment would be capable, in its current form, of being the right instrument to cure the diseased earth. He endowed the journal *GAIA—Ecological Perspectives for Science and Society*, which he helped to found at the ETH, with a notable inaugural essay, *Rethinking in the Natural Sciences* (Primas 1992):

> There is a growing recognition of the inadequacy of Cartesian and Baconian conceptions as the only basis of our understanding of nature. The aim of science is not to manipulate nature but to create insight. There are good reasons for the view that methods of contemporary science are unnecessarily limited by many preconceptions and blind fascination.

In 1992, he undertook a research excursion to Egypt together with colleagues from the newly founded department of environmental sciences, to look at how sustainable management had been practiced for thousands of years and what could be learned from this for today (Müller-Herold 2004).

# 9 Anton Amann

On 6 January 2015, Anton Amann died at the age of 58, three months after the death of his teacher. Uniquely among his PhD students, Amann had taken up Primas' scientific research program in its full diversity and developed it further. His work in theoretical chemistry, which essentially came to an end in 1996, covers elementary systems, observables, quantum logics, chirality, molecular shape, spin-boson systems, and single-molecule spectroscopy.

Anton was born as the eldest of six siblings in an entrepreneurial family in Austria. From 1974 to 1978 he studied chemistry at the ETH, where he attended Primas' lectures. As a problem for his diploma thesis, Primas gave him a hard nut to crack: What are the formal characteristics of elementary particles in algebraic quantum mechanics? This resumed earlier work by Newton and Wigner (1949), who showed that "elementarity" can be defined in relation to the kinematical symmetries of a system, i.e. to the group of its space-time symmetries. An elementary system (particle) is an object that cannot be further decomposed *by group-theoretical means*. In most cases, the relevant groups are the Lorentz and the Galilei group. As a result, it turned out that elementary particles are in one-to-one correspondence with irreducible ray representations of a kinematical group in Hilbert space. But how should one define group-related elementarity outside Hilbert space as in algebraic quantum mechanics, where symmetries are represented as automorphisms of an algebra of observables?

After a careful discussion of the relevant group-theoretical and topological arguments, Anton presented the solution of the problem: A system is elementary with respect to a group $G$ if its algebra of observables transforms *ergodically* under automorphic actions of $G$. He then showed that the known examples are special cases of his definition: Elementary systems in classical mechanics are characterized by transitive, and in Hilbert-space quantum mechanics by irreducible group actions. Primas was so puzzled that a chemistry student was able to master the mathematics with such ease that he wanted to see how far he could go and posed a second problem with an intricate computational air: to decompose a reducible representation of the Galilei group in Koopman's Hilbert space over the phase space into irreducible, i.e. elementary representations. Anton solved this in a weekend.

From January to August 1978, he carried out his military service in Tyrol, assisting a military doctor. During these eight months, he worked through a 500 p volume on mathematical foundations, Gerhard Preuss (1975) *General Topology*. In the course of this intensive self-study period, he finally developed as a mathematician. Upon returning to the ETH, Primas presented him with a mathematical proof, which he had developed himself, in which the so-called axiom of choice of transfinite set theory played a key role. Anton identified the decisive mistake. From that point onward, Primas respected the 22-year-old as a mathematical expert.

In his doctoral thesis Anton dealt with a problem of still greater complexity: not all self-adjoint operators in an operator algebra have a meaningful physical interpretation. But how can one identify the subset of observables in the strict sense, i.e. the subset of those operators which relate to well-defined physical quantities? Anton started from a time-honored example by Hermann Weyl (1927, 1928), who noticed that position and momentum operators transform differently under Galilei transformations (in modern terminology) and can thus be distinguished through their behavior under kinematical transformations. In a first step, Anton inverted Weyl's observation: observables exist only in relation to a kinematical group. "Tell me what the symmetries of your system are, and I will tell you what your observables are".

Now, every group $G$ acts in a natural way on the complex-valued group functions $f : G \rightarrow C$. He then showed that an operator is a $G$-observable if it transforms as one of those $f$. This applies to arbitrary locally compact separable groups. The traditional observables of Hilbert-space quantum mechanics fit into this scheme. Anton finally defined observables for arbitrary elementary systems: They can be constructed if a kinematical group acts via so-called *integrable* automorphisms (Amann 1986).

From 1984 to 1996, he worked initially as a research assistent and, after his *habilitation*, as a *Privatdozent* in Primas' group. Amongst others, he wrote papers on quantum logics (Amann 1987), chirality (Amann 1988), molecular shape (Amann 1993), the ground states of spin-boson systems (Amann 1991), and the quantum mechanics of single molecules (Amann 1997). He was a rising star, international recognition followed, and he received prizes and invitations for conferences, research visits, book projects and lectures.

In 1995, Primas retired, his attempts to create a permanent position for Anton having failed. Maybe he came too late, or maybe he lacked the necessary fighting spirit. Unlike Prof. Günthard, who thirty years ago overcame all institutional opposition and organized an associate professorship for him, Primas capitulated and Anton Amman had to leave the ETH and Switzerland. Where would algebraic quantum chemistry stand today if he had stayed?[8]

Initially after his departure, Anton remained in discussions with Primas about their joint research program. He voiced the suspicion that the derivation of classical observables might be mathematically too ambitious and sketched a research program of "approximately classical" states of matter (Amann and Müller-Herold 1999). In his last theoretical chemistry paper, he approached the related problem of why coherent superpositions of isomers are not observed in nature, using large deviations statistics (Amann and Loferer 2001).

## 10 The Pauli-Jung Dialog and Its Aftermath

When Kalervo Laurikainen (1988) published *Beyond the Atom: The Philosophical Thought of Wolfgang Pauli*, he introduced at a broader international level the continuing debate on Pauli's work in natural philosophy, in which the connection with C.G. Jung occupies a place of prominence. Laurikainen's book is based on the correspondence between Pauli and his former assistant Markus Fierz, at the time Professor of Theoretical Physics at Basel. Pauli took up the psychophysical problem by studying the connection between modern physics and the analytical psychology of C.G. Jung. There he encountered the concept of *unus mundus*, i.e. the idea of a primordial undivided reality from which everything emerges.

In his review for *Nature* Primas (1989) praised Laurikainen's book as a groundbreaking guide to Pauli's views. However, he found himself repeatedly disagreeing with Laurikainen and ended with the conclusion that theoretical physicists, psychologists and philosophers of science who would like to grasp Pauli's ideas stand in need of far greater help than is offered by Laurikainen's book.

---

[8]In 1996, Amann moved from the ETH to the University of Innsbruck in Austria and from theoretical to analytical chemistry. He started a medical research program on breath gas analysis, published some 150 papers on it, founded the International Association for Breath Research, served as its president, and headed the *Institut für Atemgasforschung* of the Austrian Academy of Science until his untimely death.

In the early 1990s, Primas had the good fortune to meet Harald Atmanspacher, a physicist almost thirty years younger than himself. Atmanspacher shared Primas' interest in the philosophy of science and the Pauli-Jung dialog. Together with Primas, over the following twenty years he organized lectures, edited conference reports and published more than half a dozen works in which the two developed the Pauli-Jung approach as a modern version of the dual-aspect philosophy pioneered by Baruch de Spinoza 350 years ago.

Their first common project was a symposium on the *Pauli-Jung Dialog and Its Relevance for Modern Science* at the *Monte Verità* in southern Switzerland, to which they invited natural scientists and Jungian psychologists. Primas asked me for an up-to-date account of Pauli's ideas on *teleological* explanations in molecular and evolutionary biology, which I delivered at the symposium under the title *The Sense in Chance: Reflections on Wolfgang Pauli's "Lectures to Foreign People"* (Müller-Herold 1995). It was my last commissioned work for Primas. In the dedication I wrote: "In grateful remembrance of twenty years of inimitable discussion, in which—in their own way—Jung and Pauli were a constant presence". The printed collection of the contributions enjoyed unexpected publishing success in German-speaking countries (Atmanspacher et al. 1995).

In his later years, Primas served as a member of the Pauli Committee at CERN in Geneva and as a patron of the C.G. Jung Institute at Küsnacht near Zurich. The Jung-Pauli dialog continued to be an important source of inspiration for him, particularly after the publication of the correspondence between Pauli and Jung (Meier 1992) and with Jung's coworkers Marie-Louise von Franz, Aniela Jaffé et al. (von Meyenn 1979–2005).

## 11    Grand Finale After Retirement

At the end of an article on the occasion of Primas' 70th birthday, there is a brief summarizing remark (Müller-Herold 1999, p. 8):

> Primas used to cherish the hope that the development of quantum mechanics would finally lead to results creating an impact on culture in general, beyond the bounds of natural science, and confronting philosophy, in particular, with a new situation. Couldn't the formalism bring about the downfall of the Cartesian division between mind and matter?

Just five years later, in the first issue of the journal *Mind and Matter*, Primas (2003) presented a surprising move in this direction with a paper on *Time-Entanglement between Mind and Matter*. The nature of physical time had occupied his attention for a long period, yet nothing had found its way into print. Now he made another, determined approach to the subject. His paper from 2003 begins with a literature review of the two basic concepts of time: tensed time, which relates events to the present through properties like *pastness, nowness* and *futurity*, and tenseless time or parameter time, expressed by relations like *earlier than* and *later than*.

Primas then condensed the problem of psychophysical parallelism into an interplay between tenseless time for the material sphere and tensed time for the non-material, *cum grano salis* mental sphere. His fundamental idea here, that the parallelism between mind and matter occurs without direct interaction, was not entirely new, however. It was Leibniz who presented it as a conceptual approach for the first time more than 300 years ago. He compared the psychophysical parallelism to a pair of perfectly synchronized clocks: there is no direct interaction between the clocks, the correlation is due to the initial conditions (perfect synchronization) and persists due to the—hypothetically absolute—precision of the clocks.

In quantum mechanics, correlations of non-interacting systems were introduced by Einstein, Podolsky and Rosen (1935) in an epoch-making thought experiment designed to demonstrate that quantum mechanics cannot be complete, since it admits unreasonable behavior of spatially remote subsystems. The triumph of quantum mechanics came some decades later, when Alain Aspect and his coworkers were able to demonstrate experimentally the existence of these so-called EPR-correlations of non-interacting, spatially separated quantum systems (Aspect et al. 1982). That the material world cannot always be decomposed into separable elements of reality has since then been established as a unique property of quantum systems. It is the empirical basis of quantum-mechanical holism.

Primas' approach to mind-matter correlations applied this idea in an unexpected and creative way: by replacing correlations between two material systems with correlations between a material and a non-material system. He started with a Hilbert space $\mathcal{H}$ for the undivided primordial universe of discourse without mind, matter and time. The operators on $\mathcal{H}$ do not have any extra-mathematical meaning; they are purely formal symbols. Without loss of generality, he gave $\mathcal{H}$ a tensor-product structure $\mathcal{H} = \mathcal{N} \otimes \mathcal{M}$. This was followed by a decisive technical step, which distinguishes Primas' approach from all other known attempts in this direction: To obtain interpretable quantities, he specified—in the sense of the Weyl–Wigner–Amann program—a kinematical group. The nature of this group and its presentation in the primordial Hilbert space for the *unus mundus* then determine what kind of observable facts can be investigated and stated.

As a kinematical group, he selected the extended affine group and specified a (unitary ray) representation on $\mathcal{H}$. For this choice he gave no motivation but simply mentioned that it is the relevant group for wavelet analysis. What comes next is a virtuoso application of the instruments in his mathematical tool box. It turns out that the state of the undivided universe of discourse, which is invariant under the action of the symmetry group, is a maximally time-correlated vector. Following Schrödinger (1935), who coined the notion of *entanglement* for quantum correlations, Primas denoted this as a maximally time-entangled state. Statements on the material or the non-material part can be obtained by conditional expectations from $\mathcal{H}$ onto $\mathcal{M}$ or $\mathcal{N}$ respectively.

Perhaps for didactical reasons, his exposition did not follow the logical order of concepts. He started with two groups: the so-called symmetry group of the timeless universe of discourse and the affine group. A brief inspection shows, however, that in the given representation the first group can be seen as a one-parameter subgroup of the affine group. In addition he did not(!) mention the kinematical role of the extended affine group. This was changed in an amendment to the 2003 paper (Primas 2009), where in Sect. 9.2 he even mentioned group-related elementarity.

At the end of his article, Primas wrote:

> The proposed ideas are of fragmentary and speculative character so that this essay should be considered as an exercise, whose aim is not to solve any concrete problem but to discuss new ways of thinking about the mind-matter problem.

Although this may be partly true, it is a noble understatement. There are non-trivial results:

- The construction of a time operator $T$ on $\mathcal{N}$, which allows for the derivation of tensed time including an expression for the duration of the *now*.
- In $\mathcal{N}$ there is an ahistorical part sheltering the so-called "innate" and a novelty-acquiring part where non-material "mental" phenomena are laid down.
- The extended affine group contains the time-inversion operation. The breaking of this primordial symmetry generates the arrow of time.
- *Non-interacting* (!) systems always have the same arrow of time.

If $T$ is the time operator and $\Delta T \geq 0$ its indeterminacy in the maximally time-correlated state, then $\Delta T$ is related to the duration of the *now*. In the limiting case $\Delta T = 0$ the material and the mental domain are perfectly decoupled, both mathematically and in the Cartesian sense. As in physics, there is no *now* in this case. For small $\Delta T$ synchronistic effects may occur but one does not necessarily become aware of them.

## 12 Legacies

In October 2013, I received a letter from Primas:

> I have allowed myself to revisit our considerations from 1977 in a different context. As we correctly surmised, the unpublished Chap. 7 still contained stupid errors. Nevertheless, the idea is very much a useful one and I have now revived it in the context of a draft concerning sequential and non-sequential concepts of time.

For me, this letter marked the end of a relationship that lasted more than forty years with a scientist who forged his own path independently, without asking others. Who paid close attention to research literature, but adopted nothing without painstakingly testing its validity. Who always had the creative energy to initiate new developments and to advance them further. This is, so to say, the personal part of his legacy.

On the occasion of the fiftieth anniversary of the Laboratory of Physical Chemistry in 2006 he left a legacy of a different kind to his colleagues, in favor of algebraic quantum mechanics because it enables a far simpler and clearer description of quantum systems than is allowed by the mathematical resources provided in our elementary teaching materials. (It) offers not only powerful new mathematical tools for solving classical problems, but also—and far more importantly—a new intellectual approach to research questions in the natural sciences.

Since our historical worm's-eye perspective does not allow for sound long-term prediction, I feel free to interlace an admittedly subjective statement: For me the kinematical turn in the 2003 mind-body paper is perhaps one of his most seminal contributions. Time-entanglement as a novel idea already gives the psychophysical problem a fresh perspective. The specification of kinematical symmetries then raises the discussion to a completely new level, far beyond the flimsily constructed arguments that are often apparent in the field. Maybe it will turn out that a single paper written in 2003 was the nucleus of a new stage in the development of consciousness science.

**Acknowledgments** I gratefully acknowledge linguistic help from Ursula Lindenberg (UK) and technical support by Werner Angst and Andreas Fischlin (Zurich)

# References

Amann, A. (1986): Observables in W*-algebraic quantum mechanics. *Fortschritte der Physik* **34**, 167–215.

Amann, A. (1987): Jauch-Piron states in W*-algebraic quantum mechanics. *Journal of Mathematical Physics* **28**, 2384–2389.

Amann, A. (1988): Chirality as a classical observable in algebraic quantum mechanics. In *Fractals, Quasicrystals, Chaos, Knots and Algebraic Quantum Mechanics*, ed. by A. Amann *et al.*, Dordrecht, Kluwer, pp. 305–325.

Amann, A. (1993): The gestalt problem in quantum theory: Generation of molecular shape. *Synthese* **97**, 125–156.

Amann, A. (1991): Ground states of a spin-boson model. *Annals of Physics* **208**, 414–448.

Amann, A. (1997): Quantum mechanics of single molecules. *Tatra Mountains Mathematical Publications* **10**, 159–178.

Amann, A., and Müller-Herold, U. (1999): Fragments of an algebraic quantum mechanics program for theoretical chemistry. In *On Quanta, Mind, and Matter. Hans Primas in Context*, ed. by H. Atmanspacher *et al.*, Dordrecht, Kluwer, pp. 39–51.

Amann, A., and Loferer, M.A. (2001): The problem of substance in chemistry – An approach by means of large deviation statistics. *Crystal Engineering* **4**, 101–111.

Araki, H., Haag, R., Kastler, D., and Takesaki, M. (1977): Extension of KMS states and chemical potential. *Communications in Mathematical Physics* **53**, 97–134.

Araki, H. (1983): Review of Müller-Herold (1982). *Mathematical Reviews* 83i:82011.

Araki, H. (1986): Review of Müller-Herold (1984). *Mathematical Reviews* 86d:82011.

Aspect, A., Grangier, P., and Roger, G. (1982): Experimental realization of Einstein-Podolsky-Rosen-Bohm Gedankenexperiment. A new violation of Bell's inequalities. *Physical Review Letters* **49**, 91–94.

Atmanspacher, H., Primas, H., and Wertenschlag-Birkhäuser, E., eds. (1995): *Der Pauli-Jung Dialog und seine Bedeutung für die moderne Wissenschaft*, Springer, Berlin.

Einstein, A., Podolsky, B., and Rosen, N. (1935): Can quantum-mechanical description of physical reality be considered complete? *Physical Review* **47**, 777–780.

Ernst, R.R. (1999): Hans Primas and nuclear magnetic resonance. In *On Quanta, Mind, and Matter. Hans Primas in Context*, ed. by H. Atmanspacher *et al.*, Kluwer, Dordrecht, pp. 9–38.

Feyerabend, P. (1983): Letter to Primas of October 23.

Gorini, V., Kossakowski, A., and Sudarshan, E.C.G. (1976): Completely positive dynamical semi-groups of N-level systems. *Journal of Mathematical Physics* **17**, 821–825.

Haag, R., and Kastler, D. (1964): An algebraic approach to quantum field theory. *Journal of Mathematical Physics* **5**, 848–861.

Heitler, W., and London, F. (1927): Wechselwirkung neutraler Atome und homöopolare Bindung nach der Quantenmechanik. *Zeitschrift für Physik* **44**, 455–472.

Hepp, K. (1972): Quantum theory of measurement and macroscopic observables. *Helvetica Physica Acta* **45**, 237–248.

Jauch, J.M., and Misra, B. (1961): Supersymmetries and essential observables. *Helvetica Physics Acta* **34**, 699–709.

Jörgensen, C.K. (1982): Review of H. Primas, Chemistry, Quantum Mechanics, and Reductionism. *Chimia* **36**, 221–222.

Jung, C.G., and Pauli, W. (1952): *Naturerklärung und Psyche*, Rascher, Zürich. English translation by R.F.C. Hull as *The Interpretation of Nature and the Psyche*, Routledge, London 1955.

Klein, U. (2003): *Experiments, Models, Paper tools. Cultures of Organic Chemistry in the Nineteenth Century*, Stanford University Press, Stanford.

Kolos, W., and Roothaan, C.C. (1960): Accurate electronic wave functions for the $H_2$ molecule. *Reviews of Modern Physics* **32**, 219–232.

Laurikainen, K. (1988): *Beyond the Atom: The Philosophical Thought of Wolfgang Pauli*. Springer, Berlin.

Layzer, A.J. (1963): Properties of the one-particle Green's function for non-uniform many-Fermion systems. *Physical Review* **129**, 897–907.

Lindblad, G. (1976): On generators of quantum dynamical semigroups. *Communications in Mathematical Physics* **48**, 119–130.

Lockhart, C.M., and Misra, B. (1986): Irreversibility and measurement in quantum mechanics. *Physica A* **136**, 47–76.

Longuet-Higgins, H.C. (1981): Review of H. Primas, Chemistry, Quantum Mechanics, and Reductionism. *Nature* **294**, 194.

Meier, C.A., ed. (1992): *Wolfgang Pauli und C.G. Jung. Ein Briefwechsel 1932-1958*, Springer, Berlin. English translation by D. Roscoe as *Atom and Archetype: The Pauli/Jung Letters 1932-1958*, Routledge, London 2001.

Migdal, A.B., and Larkin, A.I. (1964): Phenomenological approach to the theory of the nucleus. *Nuclear Physics* **51**, 561–582.

Müller-Herold, U. (1975): General mass-action kinetics: positiveness of concentrations as structural property of Horn's equation. *Chemical Physics Letters* **33**, 467–470.

Müller-Herold, U. (1980): Disjointness of $\beta$-KMS states with different chemical potential. *Letters in Mathematical Physics* **4**, 45–47.

Müller-Herold, U. (1982): Chemisches Potential, Reaktionssysteme und algebraische Quantenchemie. *Fortschritte der Physik* **30**, 1–73.

Müller-Herold, U. (1984): Algebraic theory of the chemical potential and the condition of reactive equilibrium. *Letters in Mathematical Physics* **8**, 127–133.

Müller-Herold, U.(1995): *Vom Sinn im Zufall: Überlegungen zu Wolfgang Paulis Vorlesung an die fremden Leute*. In *Der Pauli-Jung Dialog und seine Bedeutung für die moderne Wissenschaft*, ed. by H. Atmanspacher *et al.*, Springer, Berlin, pp. 159–177.

Müller-Herold, U. (1999): The Primas effect. In *On Quanta, Mind and Matter. Hans Primas in Context*, ed. by H. Atmanspacher *et al.*, Kluwer, Dordrecht, pp. 3–8.

Müller-Herold, U. (2004): Nachhaltigkeit – auf Altägyptisch? *GAIA* **13**, 290–291.

Müller-Herold, U. (2005): Zweifel, Hoffnung und Versagen. Junge Wissenschaftler und das desiderium visionis. *Reformatio* **54**, 49–58.

Müller-Herold, U. (2015): Are there Helium-like protonic states of individual water molecules in liquid $H_2O$? Preprint, accessible at arXiv:1512.05342

Newton, T.D., and Wigner, E.P. (1949): Localized states for elementary systems. *Reviews of Modern Physics* **21**, 400–408.

Pfeifer, P. (1980): *Chiral Molecules – A Superselection Rule Induced by the Radiation Field*. Dissertation ETH Zurich, No. 6551.

Preuss, G. (1975): *Allgemeine Topologie*, Springer, Berlin.

Primas, H. (1964): Was sind Elektronen? *Helvetica Chimica Acta* **47**, 1840–1851.

Primas, H. (1968): Zur Theorie grosser Molekeln. I. Revision der Grundlagen der Quantenchemie. *Helvetica Physics Acta* **51**, 1037–1051.

Primas, H. (1969): An exactly soluble model for the measurement process in quantum mechanics. Unpublished manuscript.

Primas, H. (1981): *Chemistry, Quantum Mechanics and Reductionism. Perspectives in Theoretical Chemistry*, Springer, Berlin.

Primas, H. (1987): Review of Lockhart and Misra (1986). *Mathematical Reviews* 87k:81006.

Primas, H. (1989): *Great expectations. Nature* **338**, 305–306.

Primas, H. (1991): Reductionism: Palaver without Precedent. In *The Problem of Reductionism in Science*, ed. by E. Agazzi, Kluwer, Dordrecht, pp. 161–172.

Primas, H. (1992): Umdenken in der Naturwissenschaft. *GAIA* **1**, 5–15.

Primas, H. (2003): Time-entanglement between mind and matter. *Mind and Matter* **1**, 81–119.

Primas, H. (2009): Complementarity of mind and matter. In *Recasting Reality. Wolfgang Pauli's Philosophical Ideas and Contemporary Science*, ed. by H. Atmanspacher and H. Primas, Springer, Berlin, pp. 171–209.

Primas, H., and Müller-Herold, U. (1978): Quantum mechanical system theory. A unifying framework for observations in quantum mechanics. *Advances in Chemical Physics* **38**, 1–107.

Primas, H., and Müller-Herold, U. (1984): *Elementare Quantenchemie*, Teubner, Stuttgart.

Raggio, G., and Primas, H. (1984): Remarks on completely positive maps in generalized quantum mechanics. *Letters in Mathematical Physics* **8**, 127–133.

Schrödinger, E. (1935): Discussion of probability relations between separated systems. *Proceedings. of the Cambridge Philosophical Society* **31**, 555–563.

Stinespring, W.F. (1955): Positive functions of C*-algebras. *Proceedings of the American Mathematical Society* **6**, 211–215.

Sutcliffe, B. (1983): Review of II. Primas, Chemistry, Quantum Mechanics, and Reductionism. *Journal of Molecuar Structure* **89**, 189–193.

von Meyenn, K., ed. (1979–2005): *Wolfgang Paulis wissenschaftlicher Briefwechsel mit Bohr, Einstein, Heisenberg u.a.*, Springer, Berlin.

von Neumann, J. (1932): *Mathematische Grundlagen der Quantenmechanik*, Springer, Berlin. English translation: *Mathematical Foundations of Quantum Mechanics*, Princeton University Press, Princeton, New Jersey, 1955.

Weyl, H. (1927): Quantenmechanik und Gruppentheorie. *Zeitschrift für Physik* **46**, 1–46.

Weyl, H. (1928): *Gruppentheorie und Quantenmechanik*, Hirzel, Leipzig. English translation: *The Theory of Groups and Quantum Mechanics*, Methuen, London, 1931; reprint: Dover, New York, 1950.

Wightman, A.S., and Glance, N. (1989): Superselection rules in molecules. *Nuclear Physics B (Proc. Suppl.)* **6**, 202–206.

# Superselection Rules

Domenico Giulini

**Abstract** Hans Primas' work on the conceptual and mathematical foundations of quantum mechanics is remarkable in many aspects. It is conceptually deeply rooted in rigorous and abstract mathematics developed by von Neumann, Mackey, Wigner, Weyl and others, and yet aims to explain the allegedly simple, like "molecules". But the classical world of localized objects in space and time is, as we know, extremely hard to reconcile with quantum mechanics. Mathematical rigor may be suspected to go along with conceptual clarity, though this is not automatically guaranteed. An important notion in this attempt to understand the quantum-to-classical transition is that of a superselection rule. In my contribution I recall some issues and developments surrounding this notion, some of which I had the pleasure to discuss with Hans Primas. I hope to show that these discussions touch upon some epistemological points of wider interest in connection with applying mathematics to the physical and sensual world.

## 1 Introduction

I first met Hans Primas in October 1992, when he and Anton Amann visited the *Forschungsstätte der Evangelischen Studiengemeinschaft e.V.* (FESt) in Heidelberg to meet with six people[1] (henceforth referred to as "we") in order to discuss the meaning and use of the notion of *decoherence* in quantum mechanics. Our group held more or less regular meetings at the FESt in order to discuss problems at the foundations of physics, in particular quantum mechanics, quantum field theory, general relativity

---

[1] These were, in alphabetical order: Domenico Giulini, Erich Joos, Claus Kiefer, Joachim Kupsch, Ion-Olimpiu Stamatescu, and H. Dieter Zeh.

D. Giulini (✉)
Institute for Theoretical Physics, University of Hannover, Hannover, Germany
e-mail: giulini@itp.uni-hannover.de

D. Giulini
Center for Applied Space Technology and Microgravity, University of Bremen,
Bremen, Germany

© Springer International Publishing Switzerland 2016
H. Atmanspacher and U. Müller-Herold (eds.), *From Chemistry to Consciousness*, DOI 10.1007/978-3-319-43573-2_4

and related topics. At that time we had a vague plan to come up with a book in the not too distant future, but that plan had then not materialized in any serious sense. The book was later published in 1996 and in its second, substantially expanded edition in 2003 (Joos et al. 2003).

In 1992 two of us (Zeh and his former Ph.D. student Joos) had already earned wide recognition as pioneers of the field of decoherence studies in quantum mechanics, mainly through the seminal papers by Zeh (1970, 1973) and Joos and Zeh (1985). The latter contained for the first time many quantitative estimates which showed beyond reasonable doubt the enormous efficiency with which environmental decoherence could suppress macroscopic superposition states to be observed. Hence a natural topic that came up in our discussions was the extent to which decoherence could *explain* the classical world around us. Claus Kiefer, also a former Ph.D. student of Zeh's, had investigated the impact of decoherence in quantum cosmology (Kiefer 1987), quantum gravity, and quantum electrodynamics (Kiefer 1992). Together we wanted to explore more of the meaning and range of applicability of these ideas to collapse models in quantum mechanics (Stamatescu) and quantum field theory (Kupsch).

My role was to understand the relation between the dynamical process of decoherence and various symmetry-based arguments of apparently purely kinematical nature which are often employed to demonstrate the existence of incoherent sectors in physical state space. Could it be that many (or perhaps all) of the symmetry arguments—mathematically rigorous as they seemed—actually relied implicitly on highly idealized dynamical assumptions of only approximate validity, and that the proper physical explanation had to take into account the dynamical interaction with the environment? After all, to state a physically meaningful symmetry of a "system" (always to be thought of as a subsystem of the world) means to state the invariance of *some* of its properties (relative to the rest of the world) with respect to *some* changes of its state (relative to the rest of the world). Pursuing this question led me deeply into some conceptual questions concerning the notion of *dynamical symmetry* that still occupies my thinking until today. From that time on I truly appreciated the remark that Wightman and Glance had made in their beautiful review of "superselection rules in molecules" (Wightman and Glance 1989):

> *The theoretical results currently available fall into two categories: rigorous results on approximate models and approximate results on realistic models.*

I always understood my discussions with Hans Primas to be essentially about the faults and virtues of these approaches. "That's a *theorem*", Primas liked to say, and sometimes I dared replying: "yes, so let's discuss its hypotheses".

Molecular superselection, chirality in particular, had been in the focus of interest of Primas, his former Ph.D. student Pfeifer (1980) and his then assistant Amann (1991), which fell more into the second class as defined by Wightman and Glance, though serious attempts were made to keep the derivations as mathematically rigorous as possible. A natural question for me was then whether those apparently strict superselection rules that followed from assumed fundamental symmetries are truly more fundamental than those approximate ones, or whether perhaps we could, and eventually need to, give a dynamical (and hence approximate) explanation for all of them. This would of course have been much welcomed by the pioneers of decoherence theory, who would not accept any fundamental inhibition to the superposition principle.

On October 17th 1992 we had our meeting with Hans Primas and Anton Amann at FESt. Primas was very well prepared and distributed a 25-p manuscript written specifically for that occasion; see Fig. 1. This manuscript is full of pointed remarks through which Primas hoped to bring home his concerns regarding the mathematical formulation and explanatory power of decoherence. He maintained that the mathematical form given to it by the pioneers was insufficient and still trapped in outdated concepts of what he called "pioneer quantum mechanics", and that the conclusions and hopes drawn by its proponents were unwarranted. After a one-page introduction he exemplified this latter point by a 10-p section "Examples of False Statements", which was not universally appreciated as a good starting point for a scientific discussion, given that many of the "false statements" were those advocated by the pioneers of decoherence theory in the audience.

In the following three Sects. 3 (7 p), 4 (3 p) and 5 (one page) Primas explained his view of the origin of the allegedly false statements. Since this also involved the *superposition principle*, which is clearly absolutely central in decoherence theory and hence very dear to its pioneers, discussions between Primas and Amann on one side and the pioneers of decoherence theory on the other soon turned nervous and further understanding soon came to a halt. As it often happens, the basic difference could be localized in the very basic first principles: Is the Schrödinger equation and the notion of *state* to be taken as basic, whereas the notion of *observable* is derived, contextual, and hence of no or little explanatory power for decoherence, as the pioneers believed? Or is the more modern algebraic formulation of quantum mechanics to be assumed as starting point, which is based on the algebra of observables to start with, from which the notion of state is derived, as Primas and Amann maintained?

I was much intrigued by this discussion and turned to the study of Primas' manuscript, which immediately guided me to his book on "Quantum Mechanics and Reductionism" (Primas 1981), whose freshness and wit I truly enjoyed. I was also much impressed by the unusual conceptual frankness and clarity of his joint book with Ulrich Müller-Herold on "elementary quantum chemistry" (Primas and Müller-Herold 1990). All these readings allured me to learn more about the "other" (algebraic) view on quantum mechanics and had the positive effect that I decided that I could learn much in trying to distil the true arguments from both sides in our discussions.

I much appreciated the mathematical setting and language of the Primas–Amann camp, but at the same time was convinced that it should not be used to kill off the ideas formulated sometimes rather intuitively in terms of concepts that presupposed a certain ontological interpretation of states in quantum mechanics (Zeh 1970). In fact, I soon found out that Primas (1981) also entertained the notion of ontic states which could be seen as a dual concept to that I had learned from the decoherence pioneers. To my mind, many of the initial apparent contradictions simply disappeared if looked at carefully and benevolently.

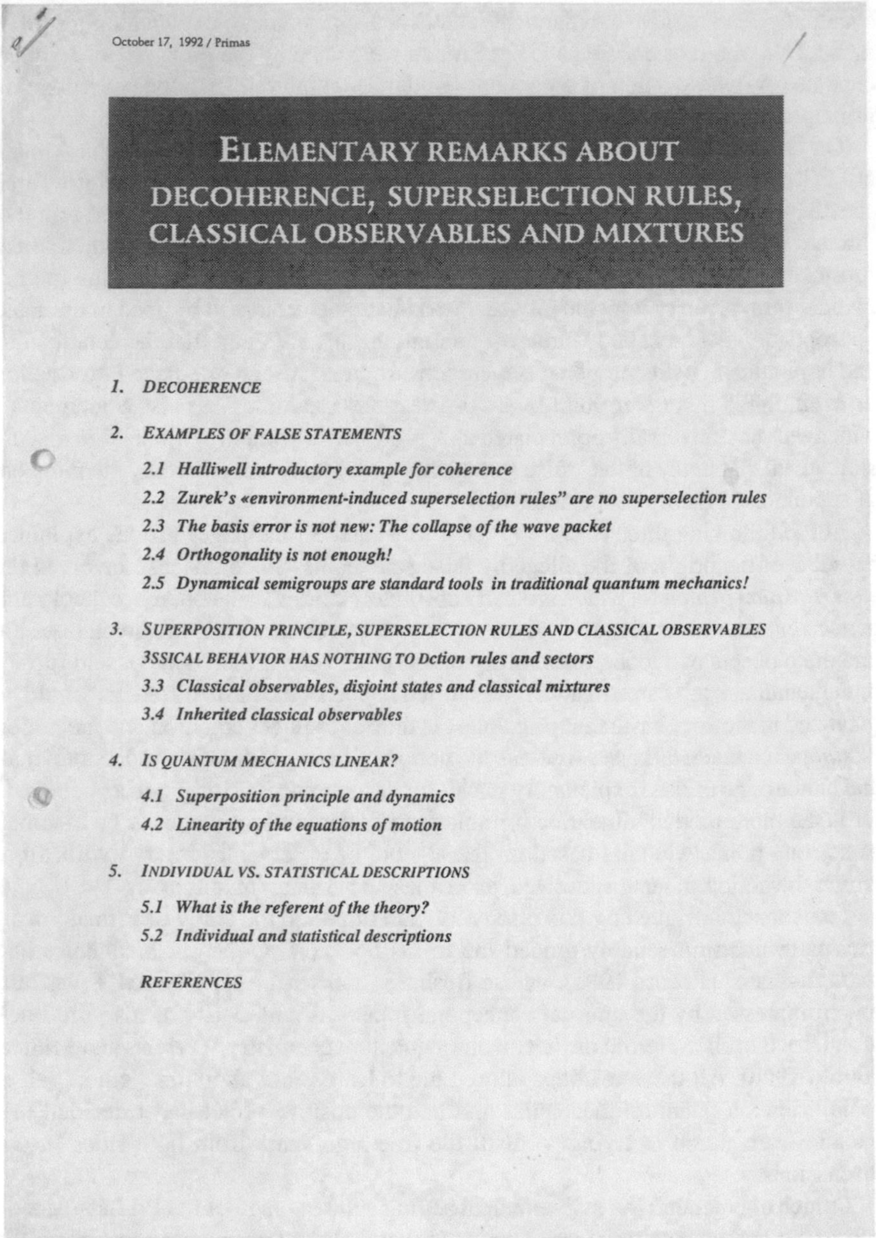

October 17, 1992 / Primas

## Elementary remarks about decoherence, superselection rules, classical observables and mixtures

**Fig. 1** Cover page of Primas' manuscript with date in the *upper left* corner

In the following sections I will take the issue of superselection rules as a guiding principle to not only recall our discussions about physics and mathematics proper, but also to illustrate the interplay between the arts of intuitive hypothesizing on one hand, and that of rigorous deduction on the other. To me the tension between these two poles of scientific thinking were the true driving force throughout all of our discussions.

# 2 Origin and General Notion

The notion of a *superselection rule* (henceforth abbreviated SSR) was introduced in 1952 by Wick (1909–1992), Wightman, and Wigner (1902–1995) in connection with the problem of consistently assigning intrinsic parity to elementary particles (Wick et al. 1952). They understood an SSR as generally expressing "restrictions on the nature and scope of possible measurements".

The concept of SSR should be contrasted with that of an ordinary *selection rule* (SR). The latter refers to a dynamical inhibition of some transition, usually due to the existence of a conserved quantity. Well known SRs in quantum mechanics concern radiative transitions of atoms. For example, in case of electric dipole radiation they take the form $\Delta J = 0, \pm 1$ (except $J = 0 \to J = 0$) and $\Delta M_J = 0, \pm 1$. It says that the quantum numbers $J$, $M_J$ associated with the atom's total angular momentum may at most change by one unit. But this is only true for electric dipole transitions, which, if allowed, represent the leading-order contribution in an approximation for wavelengths much larger than the size of the atom. The next-to-leading-order contributions are given by magnetic dipole and electric quadrupole transitions, and for the latter $\Delta J = \pm 2$ *is* possible.

This is a typical situation as regards SRs: They are valid for the leading-order modes of transition, but not necessarily for higher order ones. In contrast, a SSR is usually thought of as making a more rigorous statement. It not only forbids certain transitions through particular modes, but altogether as a matter of some deeper lying principle; hence the "super". In other words, transitions are not only inhibited for the particular dynamical evolution at hand, generated by the given Hamiltonian operator, but for all conceivable dynamical evolutions.

More precisely, two states $\psi_1$ and $\psi_2$ are separated by a SR if $\langle\psi_1|H|\psi_2\rangle = 0$ for the given Hamiltonian $H$. In case of the SR mentioned above, $H$ only contains the leading-order interaction between the radiation field and the atom, which is the electric dipole interaction. In contrast, the states are said to be separated by a SSR if $\langle\psi_1|A|\psi_2\rangle = 0$ for *all* (physically realisable) observables $A$. This means that the relative phase between $\psi_1$ and $\psi_2$ is not measurable and that coherent superpositions of $\psi_1$ and $\psi_2$ cannot be verified or prepared. It should be noted that such a statement implies that the set of (physically realizable) observables is strictly smaller than the set of all self-adjoint operators on Hilbert space. For example, $A = |\psi_1\rangle\langle\psi_2| + |\psi_2\rangle\langle\psi_1|$ is clearly self-adjoint and satisfies $\langle\psi_1|A|\psi_2\rangle \neq 0$. Hence the statement of a SSR always implies a restriction of the set of observables as compared to the set of all (bounded) self-adjoint operators on Hilbert space. In some sense, the existence of SSRs can be formulated in terms of observables alone (see below).

Since all theories work with idealizations, the issue may be raised as to whether the distinction between SR and SSR is really well founded, or whether it could, after all, be understood as a matter of degree only. For example, dynamical decoherence is known to provide a very efficient mechanism for generating apparent SSRs, without assuming their existence on a fundamental level (Zurek 2003; Joos et al. 2003).

## 3  Elementary Theory

In the most simple case of only two *superselection sectors*, a SSR can be characterized by saying that the Hilbert space $\mathcal{H}$ decomposes as a direct sum of two orthogonal subspaces, $\mathcal{H} = \mathcal{H}_1 \oplus \mathcal{H}_2$, such that under the action of each observable vectors in $\mathcal{H}_{1,2}$ are transformed into vectors in $\mathcal{H}_{1,2}$, respectively. In other words, the action of observables in Hilbert space is reducible, which implies that $\langle \psi_1 | A | \psi_2 \rangle = 0$ for each $\psi_{1,2} \in \mathcal{H}_{1,2}$ and all observables $A$. This constitutes an inhibition to the superposition principle in the following sense: Let $\psi_{1,2}$ be normed vectors and $\psi_+ = (\psi_1 + \psi_2)/\sqrt{2}$, then

$$\langle \psi_+ | A | \psi_+ \rangle = \tfrac{1}{2}\big( \langle \psi_1 | A | \psi_1 \rangle + \langle \psi_2 | A | \psi_2 \rangle \big) = \mathrm{Tr}(\rho A), \tag{1}$$

where

$$\rho = \tfrac{1}{2}\big( |\psi_1\rangle\langle\psi_1| + |\psi_2\rangle\langle\psi_2| \big). \tag{2}$$

Hence, considered as state (expectation-value functional) on the given set of observables, the density matrix $\rho$ corresponding to $\psi_+$ can be written as a nontrivial convex combination of the (pure) density matrices for $\psi_1$ and $\psi_2$ and therefore defines a mixed state rather than a pure state. Relative to the given observables, coherent superpositions of states in $\mathcal{H}_1$ with states in $\mathcal{H}_2$ do not exist or, more precisely, do not represent pure states. The set of pure states is hence *not* represented by $\mathcal{H}_1 \oplus \mathcal{H}_2$ but only by the non-linear subset $\mathcal{H}_1 \cup \mathcal{H}_2$. Vectors in the complement of that subset represent mixed states.

In direct generalization, a characterization of *discrete SSRs* can be given as follows: There exists a finite or countably infinite family $\{P_i | i \in I\}$ of mutually orthogonal ($P_i P_j = 0$ for $i \neq j$) and exhaustive ($\sum_{i \in I} P_i = 1$) projection operators ($P_i^\dagger = P_i$, $P_i^2 = P_i$) on Hilbert space $\mathcal{H}$, such that each observable commutes with all $P_i$. Equivalently, one may also say that states on the given set of observables (here represented by density matrices) commute with all $P_i$, which is equivalent to the identity

$$\rho = \sum_i P_i \rho P_i. \tag{3}$$

We define $\lambda_i := \mathrm{Tr}(\rho P_i)$ and let $I' \subset I$ be the subset of indices $i$ for which $\lambda_i \neq 0$. If we further set $\rho_i := P_i \rho P_i / \lambda_i$ for $i \in I'$, then (3) is equivalent to

$$\rho = \sum_{i \in I'} \lambda_i \rho_i, \tag{4}$$

showing that $\rho$ is a non-trivial convex combination if $I'$ contains more than one element. The only pure states are the projectors onto rays within a single $\mathcal{H}_i$. In other words, pure states are represented by vectors in the (non-linear) subset

$$\bigcup_{i \in I} \mathcal{H}_i \subset \bigoplus_{i \in I} \mathcal{H}_i. \tag{5}$$

If, conversely, *any* non-zero vector in this union defines a pure state, with different rays corresponding to different states, one speaks of an *abelian superselection rule*. In this case pure states are in bijective correspondence to rays in the subset (5). The $\mathcal{H}_i$ are then called *superselection sectors* or *coherent subspaces* on which the observables act irreducibly. The subset $Z$ of observables commuting with all observables is then given by $Z := \{\sum_i a_i P_i | a_i \in \mathbb{R}\}$. They are called *superselection* or *classical observables*.

In the general case of *continuous SSRs* $\mathcal{H}$ splits as direct integral of an uncountable set of Hilbert spaces $\mathcal{H}(\lambda)$, where $\lambda$ is an element of some measure space $\Lambda$, so that

$$\mathcal{H} = \int_\Lambda d\mu(\lambda) \mathcal{H}(\lambda), \tag{6}$$

with some measure $d\mu$ on $\Lambda$. Observables are functions $\lambda \mapsto O(\lambda)$, with $O(\lambda)$ acting on $\mathcal{H}(\lambda)$. Closed subspaces of $\mathcal{H}$ left invariant by the observables are precisely given by

$$\mathcal{H}(\Delta) = \int_\Delta d\mu(\lambda) \mathcal{H}(\lambda), \tag{7}$$

where $\Delta \subset \Lambda$ is any measurable subset of non-zero measure. In general, a single $\mathcal{H}(\lambda)$ will not be a subspace (unless the measure has discrete support at $\lambda$).

In the literature, SSRs are discussed in connection with a variety of superselection observables, most notably univalence, overall mass (in non-relativistic quantum mechanics), electric charge, baryonic and leptonic charge, and also time.

## 4 Algebraic Theory

In algebraic quantum mechanics, a system is characterized by a $C^*$-algebra $\mathfrak{C}$. Depending on contextual physical conditions, one chooses a representation

$$\pi : \mathfrak{C} \to \mathcal{B}(\mathcal{H}) \tag{8}$$

in the (von Neumann) algebra of bounded operators on Hilbert space $\mathcal{H}$. After completing the image of $\pi$ in the weak operator topology on $\mathcal{B}(\mathcal{H})$ (a procedure sometimes called *dressing* of $\mathfrak{C}$, cf. Bogolubov et al. 1990) one obtains a von Neumann sub-algebra $\mathfrak{N} \subset \mathcal{B}(\mathcal{H})$, called the *algebra of (bounded) observables*. In principle, the physical observables proper correspond to the self-adjoint elements of $\mathfrak{N}$.[2] The *commutant* $\mathcal{S}'$ of any $\mathcal{S} \subseteq \mathcal{B}(\mathcal{H})$ is defined by

$$\mathcal{S}' := \{A \in \mathcal{B}(\mathcal{H}) \mid AB = BA, \forall B \in \mathcal{S}\}, \tag{9}$$

which is automatically a von Neumann algebra. Note that the commutant is taken in $\mathcal{B}(\mathcal{H})$, i.e. comprises all elements in $\mathcal{B}(\mathcal{H})$ commuting with all elements in $\mathcal{S}$ and not just those in $\mathcal{S}$. The latter are $\mathcal{S}' \cap \mathcal{S}$. One calls $\mathcal{S}'' := (\mathcal{S}')'$ the von Neumann algebra generated by $\mathcal{S}$. It is the smallest von Neumann sub-algebra of $\mathcal{B}(\mathcal{H})$ containing $\mathcal{S}$, so that if $\mathcal{S}$ was already a von Neumann algebra one has $\mathcal{S}'' = \mathcal{S}$. The *algebra of physical observables* in the context defined by the representation (8) is then defined by

$$\mathfrak{N} := \big(\pi(\mathfrak{C})\big)''. \tag{10}$$

Let us now pause for a brief moment to clarify non-intentional implications of this terminology. To this end we quote Hans Primas (1978, p. 76):

> Traditionally the [von] Neumann algebra $\mathfrak{N}$ has the rather misleading name "algebra of observables". What really is "observable" has to come from detailed analysis of measurements and is by no means already evident from the fundamental axioms of the theory. Without a serious discussion of actual measurements, the relevance of a statement that a quantity is considered as observable remains obscure. Nevertheless, we adopt the time-honored name observable for the self-adjoint elements of $\mathfrak{N}$ without, however, accepting the frivolous claim that these observables are related in a clear cut way to measurable quantities. The algebra of observables $\mathfrak{N}$ delineates the universe of discourse of a particular subtheory.

After these cautionary remarks, we re-enter our main discussion. SSRs are said to exist iff[3] the commutant $\mathfrak{N}'$ is not trivial, that is, iff $\mathfrak{N}'$ is different from multiples of the unit operator. Projectors in $\mathfrak{N}'$ then define the sectors. *Abelian SSRs* are characterized by $\mathfrak{N}'$ being abelian. The significance of this will be explained below. $\mathfrak{N}'$ is often referred to as *gauge algebra*. Sometimes the algebra of physical observables is *defined* as the commutant of a given gauge algebra. That the gauge algebra is abelian is equivalent to $\mathfrak{N}' \subseteq \mathfrak{N}'' = \mathfrak{N}$ so that $\mathfrak{N}' = \mathfrak{N} \cap \mathfrak{N}' =: \mathfrak{N}^c$, the center of $\mathfrak{N}$. An abelian $\mathfrak{N}'$ is equivalent to *Dirac's requirement* that there should exist a complete set of commuting observables (Jauch 1960; cf. Chap. 6 of Joos et al. 2003).

In finite-dimensional Hilbert spaces Dirac's requirement is equivalent with the hypothesis that there be sufficiently many pairwise commuting self-adjoint elements

---

[2] We said "in principle" in order to hint at a possible misunderstanding in connection with this terminology, that we will comment on below just after Eq. (10).

[3] Throughout we use "iff" as abbreviation for "if and only if".

of $\mathfrak{N}$, so that the simultaneous eigenspaces are one-dimensional. In other words, each array of eigenvalues ("quantum numbers"), one for each self-adjoint element, uniquely determines a pure quantum state (a ray in $\mathcal{H}$). This implies the existence of a self-adjoint $A \in \mathfrak{N}$ with a simple spectrum (pairwise distinct eigenvalues). It then follows that any other self-adjoint $B \in \mathfrak{N}$ commuting with $A$ must be a polynomial function (of degree $n - 1$ if $n = \dim(\mathcal{H})$) of $A$ and that there exists a vector $\psi \in \mathcal{H}$ so that any other $\phi \in \mathcal{H}$ is obtained by applying a polynomial (of degree $n - 1$) in $A$ to $\psi$. The vector $\psi$ is called a cyclic vector for $\mathfrak{N}$ and may be chosen to be any vector with non-vanishing components in each simultaneous eigenspace for the complete set of commuting observables (see Chap. 6 of Joos et al. 2003 for details).

The algebraic theory allows us to translate these statements to the general situation. Here, the existence of a "complete" set of commuting observables is interpreted as existence of a "maximal" abelian subalgebra $\mathfrak{A} \subset \mathfrak{N}$. Here it is crucial that "maximal" is properly understood, namely as "maximal in $\mathcal{B}(\mathcal{H})$" and not just maximal in $\mathfrak{N}$, which would be a rather trivial requirement (given Zorn's lemma, a maximal abelian subalgebra in $\mathfrak{N}$ always exists). Now, it is easy to see that $\mathfrak{A}$ is maximal abelian (in $\mathcal{B}(\mathcal{H})$) iff it is equal to its commutant (in $\mathcal{B}(\mathcal{H})$):

$$\mathfrak{A} \text{ max. abelian} \iff \mathfrak{A} = \mathfrak{A}'. \tag{11}$$

This is true since $\mathfrak{A} \subseteq \mathfrak{A}'$ certainly holds due to $\mathfrak{A}$ being abelian. On the other hand, $\mathfrak{A} \supseteq \mathfrak{A}'$ also holds since it just expresses the maximality requirement that $\mathfrak{A}$ already contains all elements of $\mathcal{B}(\mathcal{H})$ commuting with each element of $\mathfrak{A}$.

Moreover, it can be shown that the existence of a maximal abelian subalgebra $\mathfrak{A}$ in $\mathfrak{N}$ is equivalent to $\mathfrak{N}'$ being abelian:

$$\mathfrak{A} \text{ max. abelian} \iff \mathfrak{N}' \subseteq \mathfrak{N}'' = \mathfrak{N}. \tag{12}$$

The proof of this important statement is easy enough to be reproduced here: Suppose first that $\mathfrak{A} = \mathfrak{A}'$, then $\mathfrak{N} \supseteq \mathfrak{A} = \mathfrak{A}' \supseteq \mathfrak{N}'$ and hence $\mathfrak{N}' \subseteq \mathfrak{N} = \mathfrak{N}''$, implying that $\mathfrak{N}'$ is abelian. Conversely, suppose $\mathfrak{N}'$ is abelian:

$$\mathfrak{N}' \subseteq \mathfrak{N} \qquad (\mathfrak{N}' \text{ is abelian}). \tag{13}$$

Choose an abelian subalgebra $\mathfrak{A} \subseteq \mathfrak{N}$ which is maximal in $\mathfrak{N}$:

$$\mathfrak{A} = \mathfrak{A}' \cap \mathfrak{N} \qquad (\mathfrak{A} \text{ max. abelian in } \mathfrak{N}). \tag{14}$$

As already noted above, Zorn's lemma guarantees the existence of $\mathfrak{A}$ satisfying (14). We show that $\mathfrak{A}$, albeit only required to be maximal in $\mathfrak{N}$, is in fact maximal in $\mathcal{B}(\mathcal{H})$ due to $\mathfrak{N}'$ being abelian. Indeed, since $\mathfrak{A} \subseteq \mathfrak{N}$ trivially implies $\mathfrak{N}' \subseteq \mathfrak{A}'$, we have

$$\mathfrak{N}' \overset{(13)}{=} \mathfrak{N} \cap \mathfrak{N}' \subseteq \mathfrak{N} \cap \mathfrak{A}' \overset{(14)}{=} \mathfrak{A}. \tag{15}$$

Since $\mathfrak{N}' \subseteq \mathfrak{A}$ trivially implies $\mathfrak{A}' \subseteq \mathfrak{N}$, Eq. (14) immediately leads to $\mathfrak{A} = \mathfrak{A}'$. This shows that Dirac's requirement is equivalent to the hypothesis of abelian SSRs. Another requirement equivalent to Dirac's is that there should exist a *cyclic vector* $\psi \in \mathcal{H}$ for $\mathfrak{N}$. This means that the smallest closed subspace of $\mathcal{H}$ containing $\mathfrak{N}\psi :=$ $\{A\psi \mid A \in \mathfrak{N}\}$ is $\mathcal{H}$ itself.

In quantum logic a quantum system is characterized by the lattice of propositions (corresponding to the closed subspaces, or the associated projectors, in Hilbert-space language). The subset of all propositions that are compatible with all other propositions is called the *center of the lattice*. It forms a Boolean sub-lattice. A lattice is called *irreducible* iff its center is trivial (i.e. just consists of 0, the smallest lattice element). The presence of SSRs is now characterized by a non-trivial center. Propositions in the center are sometimes called *classical*.

Finally we stress once more the contextual dependence of $\mathfrak{N}$ through the choice of representation (8). For example, even if $\mathfrak{C}$ has a trivial center, $\mathfrak{N}$ may well acquire a non-trivial center through it being mapped and weakly closed in $\mathcal{B}(\mathcal{H})$. In that sense $\mathfrak{C}$ characterizes the system "as such", $\mathfrak{N}$ merely aspects of the system in a given context. $\mathfrak{C}$ gives the ontic, $\mathfrak{N}$ the contextual or epistemic description. The clear distinction between these two descriptions—the ontic and the epistemic—had always been a theme of central importance in Hans Primas' thinking. More than once he alerted us to the dangers of terminological confusions and category mistakes. Besides the "classic" reference to Primas (1981), a more recent and lucid discussion by Atmanspacher and Primas (2003) stresses not only the *difference* but also the *relation* between ontic and epistemic descriptions.

## 5 SSRs and Conserved Additive Quantities

Let $Q$ be the operator of some charge-like quantity that behaves additively under composition of systems and also shares the property that the charge of one subsystem is independent of the state of the complementary subsystem (here we restrict attention to two subsystems). This implies that if $\mathcal{H} = \mathcal{H}_1 \otimes \mathcal{H}_2$ is the Hilbert space of the total system and $\mathcal{H}_{1,2}$ are the Hilbert spaces of the subsystems, $Q$ must be of the form $Q = Q_1 \otimes 1 + 1 \otimes Q_2$, where $Q_{1,2}$ are the charge operators of the subsystems. We also assume $Q$ to be conserved, i.e. to commute with the total Hamiltonian that generates the time evolution on $\mathcal{H}$.

It is then easy to show that a SSR for $Q$ persists under the operations of composition, decomposition, and time evolution: If the density matrices $\rho_{1,2}$ commute with $Q_{1,2}$, respectively, then, trivially, $\rho = \rho_1 \otimes \rho_2$ commutes with $Q$. Likewise, if $\rho$ (not necessarily of the form $\rho_1 \otimes \rho_2$) commutes with $Q$, then the reduced density matrices $\rho_{1,2} := \text{Tr}_{2,1}(\rho)$ (where $\text{Tr}_i$ stands for tracing over $\mathcal{H}_i$) commute with $Q_{1,2}$, respectively.

This shows that if states violating the SSR cannot be prepared initially (for whatever reason, not yet explained), they cannot be created though subsystem interactions (Wick et al. 1970). This has a direct relevance for measurement theory, since it is well known that an exact von Neumann measurement of an observable $P_1$ in subsystem 1 by subsystem 2 is possible only if $P_1$ commutes with $Q_1$, and that an approximate measurement is possible only insofar as subsystem 2 can be prepared in a superposition of $Q_2$ eigenstates (Araki and Yanase 1960).

Let us see how to prove the second to last statement for the case of discrete spectra. Let $S$ be the system to be measured, $A$ the measuring apparatus and $\mathcal{H} = \mathcal{H}_S \otimes \mathcal{H}_A$ the Hilbert space of the system plus apparatus. The charge-like quantity is represented by the operator $Q = Q_S \otimes 1 + 1 \otimes Q_A$, the observable of $S$ by $P \in \mathcal{B}(\mathcal{H}_S)$. Let $\{|s_n\rangle\} \subset \mathcal{H}_S$ be a set of normalized eigenstates for $P$ so that $P|s_n\rangle = p_n|s_n\rangle$. Let $U \in \mathcal{B}(\mathcal{H})$ be the unitary evolution operator for the von Neumann measurement and $\{|a_n\rangle\} \subset \mathcal{H}_A$ a set of normalized "pointer states" with neutral pointer-position $a_0$, so that

$$U\left(|s_n\rangle|a_0\rangle\right) = |s_n\rangle|a_n\rangle \tag{16}$$

We assume the total $Q$ to be conserved during the measurement, i.e. $[U, Q] = 0$. Clearly $\langle a_n|a_m\rangle \neq 1$ if $n \neq m$, for, otherwise, this process is not a measurement at all, since $\langle a_n|a_m\rangle = 1$ iff $|a_n\rangle = |a_n\rangle$. Let now $n \neq m$, then the following lines prove the claim:

$$
\begin{aligned}
(p_n - p_m)\langle s_n|Q_S|s_m\rangle &= (p_n - p_m)\langle s_n|\langle a_0| Q |s_m\rangle|a_0\rangle \\
&= (p_n - p_m)\langle s_n|\langle a_0| U^\dagger Q U |s_m\rangle|a_0\rangle \\
&= (p_n - p_m)\langle s_n|\langle a_n| Q_S \otimes 1 + 1 \otimes Q_A |s_m\rangle|a_m\rangle \\
&= \langle a_n|a_m\rangle (p_n - p_m)\langle s_n|Q_S|s_m\rangle. 
\end{aligned}
\tag{17}
$$

The first and fourth equality follow from $\langle s_n|s_m\rangle = 0$, the second from $[U, Q] = 0$ and unitarity of $U$, and the third from (16). Equality of the left-hand side with the last expression on the right-hand side, taking into account $\langle a_n|a_m\rangle \neq 1$, is possible iff $p_n \neq p_m$ implies $\langle s_n|Q_S|s_m\rangle = 0$, which means that $Q_S$ is reduced by (i.e. acts within) each eigenspace of $P$, which in turn implies that $Q_s$ commutes with $P$, as was to be shown.

As already indicated, the reasoning above does not explain the actual existence of SSRs in the presence of conserved additive quantities, for it does not imply anything about the *initial* non-existence of SSR-violating states. In fact, there are many additive conserved quantities, like momentum and angular momentum, for which certainly no SSRs are at work. The crucial observation here is that the latter quantities are physically always understood as *relative* to a system of reference that, by its very definition, must have certain localization properties, which exclude the total system to be in an eigenstate of *relative* (linear and angular) momenta. Similarly it was argued that one may have superpositions of relatively charged states (Aharonov and Susskind 1967a). A more complete account of this conceptually important point, including a comprehensive list of references, is given by Joos et al. (2003, Chap. 6).

# 6 SSRs and Symmetries

Symmetries in quantum mechanics are often implemented via *unitary ray represen-tations* rather than proper unitary representations (here we discard anti-unitary ray representations for simplicity). A unitary ray representation is a map $U$ from the symmetry group $G$ into the group of unitary operators on Hilbert space $\mathcal{H}$ such that the usual condition of homomorphy, $U(g_1)U(g_2) = U(g_1 g_2)$, is generalized to

$$U(g_1)U(g_2) = \omega(g_1, g_2)\, U(g_1 g_2), \tag{18}$$

where $\omega : G \times G \to \mathrm{U}(1) := \{z \in \mathbb{C} \mid |z| = 1\}$ is the so-called *multiplier* that satis-fies

$$\omega(g_1, g_2)\omega(g_1 g_2, g_3) = \omega(g_1, g_2 g_3)\omega(g_2, g_3) \tag{19}$$

for all $g_1, g_2, g_3$ in $G$, so that it ensures the property of associativity: $U(g_1)\big(U(g_2)U(g_3)\big) = \big(U(g_1)U(g_2)\big)U(g_3)$.

Note that Eq. (18) does not define $(U, \omega)$ uniquely. Any function $\alpha : G \to \mathrm{U}(1)$ allows to redefine $U \mapsto U'$ via $U'(g) := \alpha(g)U(g)$. This amounts to a redefinition $\omega \mapsto \omega'$ of multipliers given by

$$\omega'(g_1, g_2) = \frac{\alpha(g_1)\alpha(g_2)}{\alpha(g_1 g_2)}\, \omega(g_1, g_2). \tag{20}$$

Two multipliers $\omega$ and $\omega'$ are called *similar* iff (20) holds for some function $\alpha$. Similarity is an equivalence relation. A multiplier is called *trivial* iff it is similar to $\omega \equiv 1$, in which case the ray representation is, in fact, a proper representation in disguise.

Regarding SSRs, the following simple mathematical lemma is now crucial: Given unitary ray representations $U_{1,2}$ of $G$ on $\mathcal{H}_{1,2}$, respectively, with non-similar multi-pliers $\omega_{1,2}$, then no ray representation of $G$ on $\mathcal{H} = \mathcal{H}_1 \oplus \mathcal{H}_2$ exists which restricts to $U_{1,2}$ on $\mathcal{H}_{1,2}$, respectively. The proof of this lemma is almost immediate: On $\mathcal{H} = \mathcal{H}_1 \oplus \mathcal{H}_2$ we have $U_1 \oplus U_2$ satisfying $(U_1(g') \oplus U_2(g'))(U_1(g) \oplus U_2(g)) = \omega_1(g', g)U_1(g'g) \oplus \omega_2(g', g)U_2(g'g)$. If this is to define a ray representation $(U, \omega)$ of $G$ on $\mathcal{H}$ it must equal $\omega(g', g)(U_1(g'g) \oplus U_2(g'g))$ for all $g', g \in G$ for some choice of multipliers $\omega_{1,2}$ within their similarity class. This is easily seen to be possible iff $\omega_1$ and $\omega_2$ are themselves similar.

This leads to *SSRs from symmetries* in the following way: From Wigner's theorem[4] we know that symmetries in quantum mechanics correspond to unitary or anti-unitary ray representations. From this and the lemma above we conclude: If all rays in a Hilbert space $\mathcal{H}$ correspond to pure states and if $G$ acts as symmetries on that state space, then $\mathcal{H}$ cannot we written as $\mathcal{H} = \mathcal{H}_1 \oplus \mathcal{H}_2$ where $G$ acts on $\mathcal{H}_{1,2}$ via ray representations $(U_{1,2}, \omega_{1,2})$ with non-similar multipliers $\omega_{1,2}$.

However, it is important to realize that this "derivation" of SSRs rests critically on the hypothesis that it is precisely the group $G$ (which must allow for non-similar multipliers, which is a statement about its group-cohomology) of which we require a unitary ray representation. In fact, there is always a slightly larger group $\bar{G}$ that is a $U(1)$ central extension of $G$ (i.e. as Lie group it has one more dimension) which, if taken instead of $G$, would *not* give rise to SSRs. The group $\bar{G}$ is constructed as follows: As sets we have $\bar{G} \stackrel{*}{=} U(1) \times G$. The group multiplication in $U(1) \times G$ is defined as

$$(z_1, g_1)(z_2, g_2) := \big(\omega(g_1, g_2) z_1 z_2, \ g_1 g_2\big). \tag{21}$$

It is easy to show that (19) ensures associativity. The neutral element is $(1, e)$ ($e$ the neutral element in $G$) and the inverse is $(z, g)^{-1} = \big([z\omega(g, g^{-1})]^{-1}, g^{-1}\big)$. Finally, the group $U(1)$ is embedded as $\{(z, e) : z \in U(1)\} \subset \bar{G}$, which is central in $\bar{G}$ (commutes with everything). The point is now this: Given a unitary ray representation $(\omega, U)$ of $G$ on Hilbert space $\mathcal{H}$, we get a proper unitary representation $\bar{U}$ of $\bar{G}$ defined by $\bar{U}(z, g) := z \cdot U(g)$. It is immediate that $\bar{U}(1, e) = \mathrm{id}_{\mathcal{H}}$ and $\bar{U}(g_1)\bar{U}(g_2) = \bar{U}(g_1 g_2)$. But for proper representations there is clearly no obstruction to extending any two representations $\bar{U}_{1,2}$ on $\mathcal{H}_{1,2}$ to $\mathcal{H} = \mathcal{H}_1 \oplus \mathcal{H}_2$; just take $\bar{U} = \bar{U}_1 \oplus \bar{U}_2$. Often it is not even necessary to take a $U(1)$ extension. An extension by a finite subgroup suffices, as for the rotation group $SO(3)$, which just needs to be extended by $\mathbb{Z}_2$ into $SU(2)$ to transform any unitary ray representation of $SO(3)$ into a proper unitary representation of $SU(2)$. This will be discussed below.

---

[4]In modern formulation Wigner's theorem is as follows: Let $\mathcal{H}$ be a Hilbert space and $\mathbb{P}(\mathcal{H})$ its associated projective space, i.e. the set of rays in $\mathcal{H}$. We write $[v] \in \mathbb{P}(\mathcal{H})$ for the ray generated by $v \in \mathcal{H} - \{0\}$ and define the product of two rays by $([v], [w]) := |\langle v, w\rangle|/(\|v\|\|w\|)$, where $\langle \cdot, \cdot \rangle$ is the inner product in $\mathcal{H}$ and $\| \cdot \|$ its associated norm. Let $T : \mathbb{P}(\mathcal{H}) \to \mathbb{P}(\mathcal{H})$ be any map (note: continuity, injectivity, or surjectivity of $T$ are not put as hypotheses) such that for all non-zero $v, w \in \mathcal{H}$ we have $(T[v], T[w]) = ([v], [w])$. Then there exists a unitary or anti-unitary operator $U$ in $\mathcal{H}$ such that $T[v] = [Uv]$ for all $v \in \mathcal{H} - \{0\}$. The theorem is attributed to Wigner because a (weaker) form of it is stated (without full proof) in Wigner (1931). However, an earlier reference is von Neumann and Wigner (1982, footnote *** on p. 207). A standard reference for a complete proof is Bargmann (1964), assuming that $T$ is a surjection (injectivity need not be hypothesized since it immediately follows from the preservation of the ray product). The intimate connection of Wigner's theorem with the fundamental theorem of complex projective geometry was presumably first pointed out by Uhlhorn (1962). Meanwhile a number of simplifications and extensions of this theorem have been published, stressing either the algebraic or geometric aspects. Two beautiful modern proofs in this direction are Geher (2014)—dropping the hypotheses of surjectivity of $T$ and separability of $\mathcal{H}$—and Freed (2012), respectively.

## 6.1 A Closer Look at Univalence SSR

An example is given by the SSR of univalence, that is, between states of integer and half-integer spin. Here $G$ is the group $SO(3)$ of proper spatial rotations. For integer spin it is represented by proper unitary representations, for half integer spin with non-trivial multipliers. This example also serves well to illustrate the mathematical subtlety that a priori the multipliers $\omega$, and hence the redefining functions $\alpha : G \to U(1)$, need not be continuous. This potentially complicates proofs of the non-triviality of multipliers, since one needs to show the non-existence of trivializing multipliers $\alpha$ within this very large class of general functions.

In case of univalence this complication may be avoided by restriction to a finite subgroup[5]. Indeed, Eq. (20) may simply be restricted to a subgroup $G' \subset G$. If $\alpha$ existed so as to achieve (20) with $\omega'(g_1, g_2) = 1$ for all $g_{1,2} \in G$, then the restriction of $\alpha$ to $G'$ achieves the same for $G'$. Conversely, if (20) for $\omega' \equiv 1$ cannot be achieved for a subgroup $G'$ of $G$, it cannot be achieved for $G$. Now, to show the SSR for univalence it is hence sufficient to show that there exists a subgroup $G'$ of $SO(3)$ for which $\omega' \equiv 1$ is impossible. Well, take $G'$ to be Klein's "Vierergruppe" $K_4 := \{E, C_1, C_2, C_3\}$, given by the identity $E$ and the three 180-degree rotations $C_1, C_2, C_3$ about the $x$, $y$, and $z$ axis, respectively. In $SO(3)$ we have $C_a^2 = 1$ and $C_a C_b = C_b C_a = C_c$ (for $a, b, c$ pairwise distinct and in cyclic order). The two-dimensional spin 1/2 ray representation is $E \mapsto 1, C_a \mapsto \gamma_a := \exp(-i\pi\sigma_a/2) = -i\sigma_a$, where $\sigma_a$ are the Pauli matrices. Hence $\gamma_a^2 = -1$ and $\gamma_a\gamma_b = -\gamma_b\gamma_a = \gamma_c$ (for $a, b, c$ pairwise distinct and in cyclic order). This shows that

$$\omega(C_a, C_a) = -1, \quad \omega(C_a, C_b) = -\omega(C_b, C_a) = 1. \qquad (22)$$

The proof for the non-existence of $\alpha$ achieving $\omega' = 1$ in (20) is now immediate, because the fraction of $\alpha(C_a)\alpha(C_b)/\alpha(C_aC_b)$ is symmetric under exchange of $C_a$ and $C_b$ (since $K_4$ is abelian), whereas (22) shows that $\omega$ is not symmetric. But a product of a symmetric and a non-symmetric function clearly cannot equal the identity function (which is symmetric).

As already mentioned above, such derivations can and have been criticized (e.g. by Weinberg 1995 and Joos et al. 2003) for depending crucially on one's prejudice of what the symmetry group $G$ should be. No SSR will follow if instead of $G$ a larger group $\bar{G}$ is initially considered. In case of univalence we had to stick to $G = SO(3)$ to deduce a SSR. For $\bar{G}$ it suffices to take a central $\mathbb{Z}_2$-extension for $\omega$ only takes values in $\{\pm 1\} \subset U(1)$. The leads to $SU(2)$, the only non-trivial such extension of $SO(3)$. Would it then be justified to address $SU(2)$ rather than $SO(3)$ as the proper group of spatial rotations, the choice $SO(3)$ being just based on an unwarranted classical prejudice?

Interestingly, this question touches upon some deeper conceptual issues connected with the old debate on *absolute versus relative motion*. Consider, for example, a typical double-slit experiment with electrons, in which an incoming wave function

---

[5]I learned this trick from Joachim Kupsch.

$\psi$ is split into the sum $\psi = \psi_{\text{left}} + \psi_{\text{right}}$ corresponding to the spatially separated left and right slit. Then all electrons that go through the right slit are rotated by $2\pi$, whereas no action is performed on the other branch. This results in a sign change of $\psi_{\text{right}}$ (since it is of half-integer spin) and hence a change of state (ray!) corresponding to

$$\psi = \psi_{\text{left}} + \psi_{\text{right}} \; \rightarrow \; \psi' = \psi_{\text{left}} - \psi_{\text{right}}. \tag{23}$$

Indeed, for non-zero and linearly independent $\psi_{\text{left/right}}$ their sum and difference are always linearly independent. This means that a rotation of one component relative to the other results in a different overall state, or, put bluntly, $2\pi$ "rotations" do lead to observable consequences! This, in short, is the argument by Aharonov and Susskind (1967b) who conclude that consequently $SU(2)$ rather than $SO(3)$ should be addressed as rotation group and that, hence, the $SO(3)$-based argument leading to the SSR of univalence is indeed unwarranted.

A critical reply to Aharonov and Susskind (1967b) was given by Hegerfeldt and Kraus (1968), making a clear distinction between a *dynamical* rotation in real time, generated by a suitable Hamiltonian, and a *kinematical* rotation as mathematically implemented by the symmetry group of space. Only the latter, they maintain, gives rise to the univalence SSR, whereas Aharonov and Susskind argue by using the former. But is this distinction really well founded? What is, physically speaking, a "kinematical rotation", if not a suitable approximation to a real dynamical process? If that is correct, the univalence SSR is no less an approximation.

This view has also been adopted by others, like Mirman (1979). Interestingly, the relative interpretation of "rotation" has first been given for global $U(1)$ phase changes (Aharonov and Susskind 1967a), so that the charge SSR, too, may appear as a mere result of an over-idealization of the concept of symmetry. See also Mirman (1970, 1979) for a lucid summary of the relative view on motion and its consequences for the non-existence of SSRs.

## 6.2  A Closer Look at Bargmann's SSR

Another often quoted example (e.g., Primas 1981, p. 78) is the Galilei group, which is also implemented in non-relativistic quantum mechanics by non-trivial unitary ray representations whose multipliers depend on the total mass of the system and are not similar for different masses. This gives rise to what is today called the *Bargmann SSR*.[6] An elementary treatment for a single mass point is given by, e.g., Primas and Müller-Herold (1990, Exercise 3.4.4.), more generally by Giulini (1996).

---

[6]Bargmann (1954) was the first to observe that the Galilei group allowed for non-similar multipliers and that the representations furnished by the solutions to the Schrödinger equation are irreducible. He did, however, not draw the conclusion that this implies an inhibition to the superposition principle for states of different mass.

We do not go into the details of how to determine the multipliers for the ray representation of the Galilei group in quantum mechanics (see, e.g., Giulini 1996). Let us merely state that, if restricted to the abelian subgroup generated by spatial translations $\vec{a}$ and boosts $\vec{v}$, the multiplier for $g_1 = (\vec{a}_1, \vec{v}_1)$ and $g_2 = (\vec{a}_2, \vec{v}_2)$ is

$$\omega\big((\vec{a}_1, \vec{v}_1), (\vec{a}_2, \vec{v}_2)\big) = \exp\left(i\frac{M}{\hbar}\vec{v}_1 \cdot \vec{a}_2\right), \tag{24}$$

where $M$ is the overall mass of the system (sum of all masses in case of multiple-particle systems). The first thing to observe is that this multiplier is clearly not similar to the trivial one, since this expression is non-symmetric in $(g_1, g_2)$, whereas any redefinition (20) only changes the symmetric part (on the abelian subgroup considered here) and hence cannot completely remove the non-symmetric expression.[7] This also shows that multipliers for different $M$ cannot be similar, since the quotient of the multipliers for different masses is again of the form (24) with $M$ being replaced by the difference of the masses. Again, by symmetry, this cannot equal an expression symmetric in $(g_1, g_2)$. Hence overall mass defines a SSR in Galilei-invariant quantum mechanics. This is discussed in more detail by Giulini (1996) and by Joos et al. (2003, Appendix A6). Again the trick is to restrict attention to a suitable abelian subgroup.

More important for our conceptual discussion here is once more the question of whether this derivation makes physical sense. Why should it not? Well, because the statement that superpositions of states with different overall mass cannot exist (as pure states) only makes sense for systems which can—in principle—assume states with different overall mass. In other words, such statements only make sense in a dynamical framework in which mass is a dynamical parameter. However, in ordinary non-relativistic quantum mechanics mass is a parameter that characterizes a system, it is not a dynamical variable. Systems with different masses are considered different systems. There is no superposition principle for states of different systems since they live in different Hilbert spaces.

The Bargmann SSR is based on the hypothesis that the Galilei group acts by unitary ray representations on physical states. Again this may be an unwarranted requirement in view of the fact that we *first* have to find a framework in which mass becomes a dynamical variable and *then* decide on its dynamical symmetries. A naive classical test theory in which $N$ masses become dynamical has been given by Giulini (1996). It is fully characterized by the action

$$S\big[\{\lambda_a\}, \{\vec{x}_a\}; \{m_a\}, \{\vec{p}_a\}\big] = \int dt \left\{\sum_a m_a \dot{\lambda}_a + \vec{p}_a \cdot \dot{\vec{x}}_a - H(\{m_a\}, \{\vec{x}_a\}, \{\vec{p}_a\})\right\}, \tag{25a}$$

---

[7]Writing $g_a = (\vec{a}_a, \vec{v}_a)$ and using $\alpha(g_a) = \exp\big[(i/2\hbar)\vec{a}_a \cdot \vec{v}_a\big]$ ($a = 1, 2$), a redefinition of the form (20) removes the symmetric part of (24) and leaves us with the anti-symmetric part:

$$\omega'\big((\vec{a}_1, \vec{v}_1), (\vec{a}_2, \vec{v}_2)\big) = \exp\left(i\frac{M}{2\hbar}(\vec{v}_1 \cdot \vec{a}_2 - \vec{v}_2 \cdot \vec{a}_1)\right).$$

where, for definiteness, we may take the $\lambda_a$-independent Hamiltonian to be of the form

$$H\{(\lambda_a), \{m_a\}, \{\vec{x}_a\}, \{\vec{p}_a\}\} = \sum_{a=1}^{N} \frac{\|\vec{p}_a\|^2}{2m_a} + V(\{m_a\}, \{\vec{x}_a\}). \qquad (25b)$$

We used the abbreviation $\{m_a\} := (m_1, \ldots, m_N)$. The $N$ new pairs of conjugate variables are $(\lambda_a, m_a)$, with $\lambda_a$ the canonically conjugate "generalized position" with respect to $m_a$. Both $\lambda_a$ and $m_a$ are dynamical variables, chosen such that the new "momenta" $m_a$ stay constant as a result of the equations of motion. Indeed, the equations of motion for the old variables are as before,

$$\dot{\vec{x}}_a = \partial H / \partial \vec{p}_a, \qquad (26a)$$

$$\dot{\vec{p}}_a = -\partial H / \partial \vec{x}_a, \qquad (26b)$$

whereas for the new ones we obtain

$$\dot{\lambda}_a = \partial V / \partial m_a - \|\vec{p}_a\|^2 / 2m_a^2, \qquad (27a)$$

$$\dot{m}_a = -\partial H / \partial \lambda_a = 0 \Rightarrow m_a = \text{const}. \qquad (27b)$$

The $m_a$ are now integration constants. Inserting them into (26) it becomes an autonomous subsystem (the same as before adjoining the new variables). Having obtained a solution $\{\vec{x}_a(t), \vec{p}_a(t)\}$ we can finally insert this into (27a) which can then be solved by quadrature.

The whole point of this exercise is the following: Once we have enlarged the phase space so as to include the masses $m_a$ as dynamical variables, we may ask for the dynamical symmetry group of the total system of equations of motion consisting of the old (26) and new (27) equations. It can then be shown that, if before adjoining $\{\lambda_a, m_a\}$ it had been the *Galilei group* for the system (26) (which is, e.g., the case if $V$ only depends on the distances $\|\vec{x}_a - \vec{x}_b\|$), the total system will have the 11-dimensional *Schrödinger group* as dynamical symmetries, which is a central $\mathbb{R}$-extension of the Galilei group that does *not* give rise to any SSR via simple symmetry requirements.

This is discussed in detail by Giulini (1996). Admittedly, this model is naive and has obvious deficiencies, like a singular Hamiltonian for vanishing masses; masses should be strictly positive. But this can be repaired without changing the conclusion. The model just presented has recently been much improved through a more rigorous formulation in terms of algebraic quantum mechanics (i.e. in a form preferred by Primas) in which the mass spectrum is positive and discrete (Annigoni and Moretti 2012). This paves the way for a proper *dynamical* understanding of a SSR for mass, which is possible if the spectrum for mass is properly discrete, i.e. with positive lower bound for all mass differences (for details see Annigoni and Moretti 2012).

# 7 SSRs in Local Quantum Field Theories

In quantum field theory, SSRs emerge through the requirements of locality and causality: $\mathfrak{N}$ is given by the (weak closure of) observables localized in space and time (smearing functions of compact support). Charges which can be measured by fluxes through closed surfaces at arbitrarily large spatial distances must then commute with all observables.

A typical example is the total electric charge, which is given by the integral over space of the local charge density $\rho$. According to Maxwell's equations, the latter equals the divergence of the electric field $\vec{E}$, so that Gauss' theorem allows us to write

$$Q = \lim_{R \mapsto \infty} \int_{S_R} (\vec{n} \cdot \vec{E}) d^2\sigma, \tag{28}$$

where $\vec{n}$ is the normal vector to the sphere $S_R := \{\vec{x} : \|\vec{x}\| = R\}$ which bounds the ball $B_R := \{\vec{x} : \|\vec{x}\| \leq R\}$, and $d^2\sigma$ its surface measure. If $A$ is a local observable, its support is in the *causal complement* of the spheres $S_R$ for sufficiently large $R$. The quantum version of (28) should then read like this:

$$\langle \Psi \,|\, [A,Q] \,|\, \Phi \rangle = \lim_{R \to \infty} \left\langle \Psi \,\middle|\, \int_{B_R} d^3x \; [A,\rho(\vec{x})] \,\middle|\, \Phi \right\rangle$$

$$= \lim_{R \to \infty} \left\langle \Psi \,\middle|\, \int_{S_R} d^2\sigma \; [A,\vec{E}(\vec{x})] \cdot \vec{n} \,\middle|\, \Phi \right\rangle, \tag{29}$$

where $\Psi$ and $\Phi$ are two states. Since in the second integral over the two-sphere $S_R$ the radius $R$ can be chosen large enough so that all of its points $\vec{x}$ lie in the causal complement of $A$, the commutator $[A,\vec{E}(\vec{x})]$ should vanish and so should all matrix elements $\langle \Psi \,|\, [A,Q] \,|\, \Phi \rangle$. Clearly, there are formal gaps in this argument, like the invalidity of Gauss' theorem as an operator identity. But it has been shown by Strocchi and Wightman (1974, 1976) that these difficulties can be overcome and that the conclusions given above are indeed valid.

If presented in this form, the charge SSR is a mathematical theorem and, as such, indisputable (as Primas liked to stress). But it rests on hypotheses whose physical validity may well be questioned (as Primas would certainly agree). The type of concern that comes to mind is somewhat similar to the one voiced by John Bell (1987) in connection with Klaus Hepp's (1972) dynamical modeling of state reduction, where "states" are taken with respect to the restricted set of *quasi-local* observables (on an infinite one-dimensional lattice of spin 1/2 systems). Observables are restricted to be of compact support, i.e. to "see" only finitely many lattice sites. But, clearly, time-dependent observables could be chosen so as to "run after" the travelling correlations, thereby never "loosing sight". The field-theoretic argument given above also crucially relies on the restriction to quasi-local observables and we may well ask for the physical basis of that restriction. Is it mandatory in any sense, or merely plausible? Before we briefly turn to that question, let us for the sake of completeness add a few more general remarks on SSRs in quantum field theory.

In modern local quantum field theory (Haag 1996), representations of the quasi-local algebra of observables are constructed through the choice of a preferred state on that algebra (GNS-construction), like the Poincaré invariant vacuum state, giving rise to the *vacuum sector*. The superselection structure is restricted by putting certain selection conditions on such states, like e.g. the Doplicher-Haag-Roberts selection criterion for theories with mass gap (there are various generalizations, see Haag 1996), according to which any representation should be unitarily equivalent to the vacuum representation when restricted to observables whose support lies in the causal complement of a sufficiently large (causally complete) bounded region in spacetime. Interestingly this can be closely related to the existence of gauge groups whose equivalence classes of irreducible unitary representations faithfully label the superselection sectors. Recently, a systematic study of SSRs in "locally covariant quantum field theory" was started by Brunetti and Ruzzi (2007). Finally we mention that SSRs may also arise as a consequence of non-trivial spacetime topology (Ashtekar and Sen 1980).

The foregoing argument that leads to the charge SSR seems to suggest an abundance of SSRs in field theory, one for each Gauss-like law. For example, in general relativity, the Poincaré charges *mass, linear, and angular momentum* are all given by surface integrals over 2-spheres at spacelike infinity:

$$
\begin{aligned}
m &= \lim_{R \to \infty} \left\{ \frac{c^2}{16\pi G} \int_{S_R} d^2\sigma \, n^a \left( \partial_b g_{ab} - \partial_a g_{bb} \right) \right\} \\
p_\xi &= \lim_{R \to \infty} \left\{ \frac{c^2}{8\pi G} \int_{S_R} d^2\sigma \, n^a \left( K_{ab} - \delta_{ab} K_{cc} \right) \xi^b \right\}
\end{aligned}
\tag{30}
$$

Here $g_{ab}$, $K_{ab}$ denote the components of the first and second fundamental form (Riemannian metric and extrinsic curvature) of the spatial (3-dimensional) Cauchy surface and $\xi$ denotes a vector field that either generates an asymptotic translation (in which case $p_\xi$ gives the linear momentum in $\xi$ direction) or an asymptotic rotation (in which case $p_\xi$ gives the corresponding angular momentum).

In this context the restriction to local observables seems less well justified. For example, an observable not commuting with angular momentum would be the spatial orientation relative to a background reference frame obtained by "looking at fixed stars", e.g., by having observable access to the *extra-galactic celestial reference frame*. This is what we usually assume in our description of quasi-isolated systems in general relativity and which is also suggested by the formalism, because asymptotic rotations and translations are *not* to be considered as genuine gauge transformations in the sense of relating redundant mathematical descriptions of the *same* physical state. Gauge transformations are characterized by the property to yield a zero action if performed in real time. But this means that the corresponding charge (which is the functional derivative of the action with respect to the coordinate gauge motion) has to vanish.

Interestingly, the latter argument also applies to the charge SSR. A proper variational formulation of Maxwell's equation to the effect that charged states (i.e. long-
ranging fields) in the domain of differentiability of the action functional must contain
surface integrals at infinity in the action that assume non-zero values if evaluated on
asymptotically non-trivial $U(1)$ transformations. Therefore, a global $U(1)$ phase
change cannot a priori be declared unobservable, unless further restrictions on the
set of observables are introduced *by hand*, like that of quasi-locality. This puts the
question on the physical limitations of a charge SSR back on the agenda. More details
on this can be found in Joos et al. (2003, Sect. 6.4.1).

## 8   Environmentally Induced SSRs

The ubiquitous mechanism of decoherence effectively restricts the *local* verifica-
tion of coherences (Joos et al. 2003). For example, scattering of light on a particle
undergoing a two-slit experiment *delocalizes* the relative phase information for the
two beams along with the escaping light. Hence effective SSRs emerge locally in a
practically irreversible manner, albeit the correlations are actually never destroyed
but merely delocalized.

The emergence of effective SSRs through the dynamical process of decoherence
has also been called *einselection* (Zurek 2003). For example, this idea has been
applied to the problem of why certain molecules naturally occur in eigenstates of
chirality rather than energy and parity, i.e. why sectors of different chirality seem to
be superselected so that chirality becomes a classical observable. This is just a special
case of the general question of how classical behavior can emerge in quantum theory. It may be asked whether *all* SSRs are eventually of this dynamically emergent
nature, or whether strictly fundamental SSRs persist on a kinematical level (Joos
et al. 2003). The complementary situation in theoretical modeling may be characterized as follows: Derivations of SSRs from axiomatic formalisms lead to exact results
on models of only approximate validity, whereas the dynamical approach leads to
approximate results on more realistic models.

## 9   SSRs in Quantum Information

In the theory of quantum information, a somewhat softer variant of SSRs is defined
to be a restriction on the allowed local operations (completely positive and trace-
preserving maps on density matrices) on a system (Bartlett and Wiseman 2003).
In general, this therefore leads to constraints on (bipartite) entanglement. Here the
restrictions considered are usually not thought of as being of any fundamental nature,
but rather for mere practical reasons. For example, without an external reference
system for the definition of an overall spatial orientation, only "rotationally covariant"
operations $\mathcal{O} : \rho \mapsto \mathcal{O}(\rho)$ are allowed, which means that $\mathcal{O}$ must satisfy

$$\mathcal{O}\left[U(g)\rho U^{\dagger}(g)\right] = U(g)\mathcal{O}(\rho)U^{\dagger}(g) \quad \forall g \in SO(3), \tag{31}$$

where $U$ is the unitary representation of the group $SO(3)$ of spatial rotations in Hilbert space. Insofar as the local situation is concerned, this may be rephrased in terms of the original setting of SSRs, e.g. by regarding $SO(3)$ as a gauge group, restricting local observables and states to those commuting with $SO(3)$. On the other hand, one also wishes to consider situations in which, for example, a local bipartite system (Alice and Bob) is given a state that has been prepared by a third party that is *not* subject to the SSR.

## 10 Conclusion

*The theoretical results currently available fall into two categories: rigorous results on approximate models and approximate results on realistic models.* This was the motto presented at the beginning, whose faults and virtues guided our (actual and sometimes also imaginative) discussions with Hans Primas. What have we learned from these discussions? Personally I think I have learned an important but, retrospectively speaking, simple lesson: The degree of rigor in mathematical reasoning applied to physics (or any other science) should not only be measured by the degree to which mathematical notions are properly defined and derivations are complete and comprehensibly connected to the hypotheses. It should also be measured by the degree to which the existence, content, and contingency of physical hypotheses are made explicit and visible. Physical inputs should be rigorously disclosed instead of being dressed up as mathematical necessity. For that, rigor in the first and obvious sense is a necessary but far from sufficient condition to be required.

Let us once more look at one of our concrete examples. In Sect. 6.2 we saw how Bargmann's SSR followed from a group-theoretic argument based on the hypothesis that state spaces would universally support certain symmetry operations. Primas was much impressed with this argument, as one can tell from the discussions in Primas (1981) and Primas and Müller-Herold (1990), and as I have heard from him personally. This is indeed a good example to be concrete, since it shows Primas' enormous confidence in the epistemic significance of mathematical reasoning. Amongst his comments on the Bargmann SSR, he wrote (Primas 1981, p. 73):

> Bargmann's superselection rule leads for the first time to a deeper understanding of the role of mass in mechanics. Bargmann's superselection rule says that the mass of an elementary particle in a Galilei invariant theory is a classical observable. That is, the Galilei group gives the final explanation of the concept of the conservation of matter introduced into chemistry by Antoine Laurent de Lavoisier (1743–1794) and of the law of definite proportions due to Joseph Louis Proust (1754–1826) and John Dalton (1766–1844).

This is a strikingly far-reaching statement and, physically speaking, surprisingly daring in view of the fact that it is entirely based on comparatively simple mathematics associated to the structural analysis of the Galilei group. We recall that in that analysis, the physical concept of mass, conceptually elusive and multifaceted as it may

otherwise appear (Jammer 1999, 2010), degenerates to a mere parameter that labels
the inequivalent central extensions.

But can that be? Can we really *understand* the classicality and conservation of
mass in Galilean relativity by means of a "final explanation" through simple group
theory? Would it, physically speaking, not be much more appropriate to consider
a formalism less rigid and amenable to deformations through which the sought for
result appears as approximate (so as to also allow for estimates on upper bounds on
possible violations of, say, mass conservation) rather than as incontrovertible result
of a tightly tailored mathematical framework? What precisely does it take to turn
a mere mathematical model of an allegedly empirical fact into a proper or "final"
explanation? At such epistemological points discussions typically diverge, as they
did in our discussions in Heidelberg mentioned in the introduction.

As we have discussed in some detail in Sect. 6.2, if mass is considered to be a
dynamical variable, added together with its canonically conjugate variable to the
classical phase space, the structure of the ensuing dynamical symmetry group is that
of a central extension of the Galilei group, which now does *not* give rise to any SSR. So
it seems that, if possible consequences regarding SSRs depend so delicately on what
exactly the dynamical symmetries are, we should not overemphasize their physical
relevance. They are merely *"... rigorous results on approximate models"*—"rigorous"
because they are truly corollaries to theorems in group theory, "approximate" because
they certainly rest on physical assumptions of certain contingency, last not least that
the symmetry group is the Galilei group, which may physically only be approximately
true (as we know it is from special relativity).

Here Primas' emphasis seems distinctly different. For him the Galilei group stands
not as an example for a contingent dynamical symmetry that a system may or may
not share. Rather, he regards it as the *automorphism group* of spacetime (of classical
and quantum mechanics) that exists prior and independent of its material content. As
such, Primas emphasizes, it must be reflected as *kinematical symmetry group* on any
physical systems within that spacetime, which is a statement not about the *dynamics*
of the system, but rather about its *observables*. Explicitly he explains (Primas 1978,
p. 73):

> The kinematical group determines the pattern of feasible abstract motions apart from con-
> siderations of mass and force, and is therefore conceptually independent from the dynamical
> laws and the symmetries of the Hamiltonian.

This sounds as if the kinematical group was given once and for all, and hence all
its implications including possible SSRs. On the other hand, if observables are characterized by the kinematical group, there seems to be an unnecessary and moreover unphysical rigidity in not allowing the kinematical symmetry group to adapt
to one's description of a system, e.g., by adding hitherto neglected observables or
dynamical—and hence observable—degrees of freedom. No element of contextuality seems to be present.

For a long time I misunderstood Primas' approach by thinking that it could not
accommodate such a flexibility. But I was wrong. I was mislead by his references to
Hermann Weyl's group theoretic approach to observables (Weyl 1949, 1981), which

Primas sharply distinguished from and rated much higher than mere dynamical considerations à la Wigner. But it has to be taken into account that Weyl relates his approach to a kind of "relativity principle" that distinguishes *physical* and *geometric* automorphisms, and that—perhaps surprisingly—the latter are derived from the former instead of the other way around, a fact that I did not appreciate in reading Primas' representation of Weyl. More precisely, in Weyl's terminology, the geometric automorphisms are defined to be the normal closure (normalizer) of the physical automorphisms. For this to make sense we first need to know what the latter are and also in what ambient group the process of normal closure is to be taken. Weyl, it seems to me, was well aware that this refers to the ambient physical universe and its contingent dynamical laws; see Weyl (1949) and the discussions in Weyl (2016). This is clearly a contextual element in the definition of geometric symmetry that has a certain resemblance to defining the algebra of observables by the weak closure (10). And this is also how I understand Primas' reference to Weyl now.[8]

Based upon this understanding it is clear that the generalization of the concept of kinematical groups to subsystems allows them to acquire contextual features by which these *local* groups may well differ from a somehow distinguished *global* one. We have seen examples of that sort in our discussions of Bargmann's SSR, and also the SSR of univalence, where the global kinematical group of rotations, $SO(3)$ (a subgroup of the Galilei group) emphasized by Hegerfeld and Kraus (1968), differs from the local kinematical group relating relative orientations of subsystems, $SU(2)$—emphasized by Aharonov and Susskind (1967b) and Mirman (Mirman 1970, 1979). In fact, we have seen that the proper determination of local kinematical symmetries depends crucially on the dynamical interaction of the local subsystem in question with its physical environment, so that insisting on labelling it "kinematical" seems rather improper.

Moreover, since anything in the physical world interacts with its environment—except the "universe as a whole", if that notion makes physical sense at all—, one may well ask whether the notion of a truly dynamically independent kinematical group makes any deeper sense at all. In my understanding Primas had an ambivalent attitude that I could never resolve: On the one hand he clearly saw the inevitable contextuality of structural assignments (like an algebra of observables), but on the other hand he also felt the need to separate off absolute elements in order to serve the apparent needs of any decent ontology. This ambivalence shows in his surprisingly

---

[8]Another recent context in which Primas uses the concept of kinematical groups in order to define fundamental observables without initial reference to any dynamical interaction with the environment is his exploration into the phenomenology of time (Primas 2009). There he more explicitly discusses contextually broken symmetries and the need of symmetry breaking for the "creation" of phenomena. He cites Pierre Curie (1894): "C'est la dissymétrie qui crée le phénomène".

"relativistic" answer to the question of how, finally, we should decide on the "correct" kinematical group (which, recall, we had learned from him to be a priori due to dynamical considerations). Primas (1978, p. 85):

> The study of kinematical groups in quantum mechanics is the same as the study of the nature of the algebra of observables of quantal systems. [...] The question of how one has to choose the correct kinematical group is adequately answered by the Cheshire Cat in Lewis Caroll's *Alice in Wonderland*: "Would you tell me, please, which way I ought to go from here?" "That depends a good deal on where you want to get to", said the Cat. "I don't much care where", said Alice. "Then it doesn't matter which way you go", said the Cat. "– So long as I get somewhere", Alice added as an explanation. "Oh, you're sure to do that", said the Cat, "if you only walk long enough".

# References

Aharonov, Y., and Susskind, L. (1967a): Charge superselection rule. *Physical Review* **155**, 1428–1431.

Aharonov, Y., and Susskind, L. (1967b): Observability of the sign change of spinors under $2\pi$ rotations. *Physical Review* **158**, 1237–1238.

Amann, A. (1991): Chirality: A superselection rule generated by the molecular environment? *Journal of Mathematical Chemistry* **6**, 1–15.

Annigoni, E., and.Moretti, V. (2012): Mass operator and dynamical implementation of mass superselection rule. *Annales Henri Poincaré* **14**, 893–924.

Araki, H., and .Yanase, M.M. (1960): Measurement of quantum mechanical operators. *Physical Review* **120**, 622–626.

Ashtekar, A., and Sen, A. (1980): On the role of space-time topology in quantum phenomena: Superselection of charge and emergence of nontrivial vacua. *Journal of Mathematical Physics* **21**, 526–533.

Atmanspacher, H., and Primas, H. (2003): Epistemic and ontic quantum realities. In *Time, Quantum and Information*, ed. by L. Castell and O. Ischebeck, Springer, Berlin, pp. 301–321.

Bargmann, V. (1954): On unitary ray representations of continuous groups. *Annals of Mathematics (second series)* **59**, 1–46.

Bargmann, V. (1964): Note on Wigner's theorem on symmetry operations. *Journal of Mathematical Physics* **5**, 862–868.

Bartlett, S.D., and Wiseman, H.M. (2003): Entanglement constrained by superselection rules. *Physical Review Letters* **91**, 097903.

Bell, J. (1987):. On wave packet reduction in the Coleman-Hepp model. In *Speakable and Unspeakable in Quantum Mechanics*, Cambridge University Pressm Cambridge, pp. 45–51.

Bogolubov, N.N., Logunov, A.A., Oksak, A.I., and Todorov, I.T. (1990): *General Principles of Quantum Field Theory*, Kluwer, Dordrecht.

Brunetti, R., and Ruzzi, G. (2007): Superselection sectors and general covariance. I. *Communications in Mathematical Physics* **270**, 69–108.

Curie, P. (1894): Sur la symétrie dans les phénomènes physiques. *Journal de Physique* **3**: 393–416.

Freed, D.S. (2012): On Wigner's theorem. *Geometry & Topology Monographs* **18**, 83–89.

Geher, G.-P. (2014): An elementary proof for the non-bijective version of Wigner's theorem. *Physics Letters A* **378**, 2054–2057.

Giulini, D. (1996): On Galilei invariance in quantum mechanics and the Bargmann superselection rule. *Annals of Physics (New York)* **249**, 222–235.

Haag, R. (1996): *Local Quantum Physics: Fields, Particles, Algebras*, Springer, Berlin.

Hegerfeldt, G., and Kraus, K. (1968): Critical remark on the observability of the sign change of spinors under $2\pi$ rotations. *Physical Review* **170**, 1185–1186.

Hepp, K. (1972): Quantum theory of measurement and macroscopic observables. *Helvetica Physica Acta* **45**, 237–248.

Jammer, M. (2010): *Concepts of Mass in Classical and Modern Physics*. Dover, New York. Reprint of the 1961 first edition by Harvard University Press, Cambridge.

Jammer, M. (1999): *The Concept of Mass in Contemporary Physics and Philosophy*, Princeton University Press, Princeton.

Jauch, J.M. (1960): Systems of observables in quantum mechanics. *Helvetica Physica Acta* **33**, 711–726.

Joos, E., and Zeh, H.-D. (1985): The emergence of classical properties through interaction with the environment. *Zeitschrift für Physik B* **59**, 223–243.

Joos, E., Zeh, H.-D., Kiefer, C., Giulini, D., Kupsch, J., and Stamatescu, I.-O. (2003); *Decoherence and the Appearence of a Classical World in Quantum Theory*, Springer, Berlin.

Kiefer, C. (1987): Continuous measurement of mini-superspace variables by higher multipoles. *Classical and Quantum Gravity* **4**, 1369–1382.

Kiefer, C. (1992): Decoherence in quantum electrodynamics and quantum gravity. *Physical Review D* **46**, 1658–1670.

Mirman, R. (1970): Analysis of the experimental meaning of coherent superposition and the nonexistence of superselection rules. *Physical Review D* **1**, 3349–3363.

Mirman, R. (1979): Nonexistence of superselection rules: Definition of term *Frame of Reference*. *Foundations of Physics* **9**, 283–299.

Pfeifer, P. (1980): *Chiral Molecules - a Superselection Rule Induced by the Radiation Field*. PhD thesis, Swiss Federal Institute of Technology (ETH) Zürich, Diss. ETH No. 6551.

Primas, H. (1978): Kinematical symmetries in molecular quantum mechanics. In P. Kramer and A. Rieckers, editors, *Group Theoretical Methods in Physics*, ed. by R. Kramer abd A. Rieckers, Springer, Berlin, pp. 72–91.

Primas, H. (1981): *Chemistry, Quantum Mechanics and Reductionism. Perspectives in Theoretical Chemistry*, Springer, Berlin.

Primas, H. (2009): Complementarity of mind and matter. In *Recasting Reality. Wolfgang Pauli's Philosophical Ideas and Contemporary Science*, ed. by H. Atmanspacher and H. Primas, Springer, Berlin, pp. 171–209.

Primas, H., and Müller-Herold, U. (1990). *Elementare Quantenchemie*, Teubner, Stuttgart.

Strocchi, F., and Wightman, A.S. (1974): Proof of the charge superselection rule in local relativistic quantum field theory. *Journal of Mathematical Physics* **15**, 2198–2224.

Strocchi, F., and Wightman, A.S. (1976): Erratum: Proof of the charge superselection rule in local relativistic quantum field theory. *Journal of Mathematical Physics* **17**, 1930–1931.

Uhlhorn, U. (1962): Representation of symmetry transformations in quantum mechanics. *Arkiv för Fysik* **23**, 307–340.

von Neumann, J., and.Wigner, E.P. (1928): Zur Erklärung einiger Eigenschaften der Spektren aus der Quantenmechanik des Drehelektrons. *Zeitschrift für Physik* **47**, 203–220.

Weinberg, S. (1995): *The Quantum Theory of Fields. Volume I Foundations*, Cambridge University Press, Cambridge.

Weyl, H. (1949):. *Philosophy of Mathematics and Natural Science*. Princeton University Press, Princeton. Revised and augmented English edition based on a translation by Olaf Helmer of the 1927 German edition.

Weyl, H. (1981): *Gruppentheorie und Quantenmechanik*, Wissenschaftliche Buchgesellschaft, Darmstadt.

Weyl, H. (2016): *Symmetrie*, Springer, Berlin. Republication of the 1955 German edition, edited and commented by Domenico Giulini, Erhard Scholz, and Klaus Volkert, with a hitherto unpublished manuscript *Symmetry and Congruence* by Hermann Weyl.

Wick, G.C., Wightman, A.S., and Wigner E.P. (1952): The intrinsic parity of elementary particles. *Physical Review* **88**, 101–105.

Wick, G.C., Wightman, A.S., and Wigner, E.P. (1970): Superselection rule for charge. *Physical Review D* **1**, 3267–3269.

Wightman, A.S., and Glance, N. (1989): Superselection rules in molecules. *Nuclear Physics B* **6**, 202–206.

Wigner, E.P. (1931): *Gruppentheorie und ihre Anwendung auf die Quantenmechanik der Atomspektren*, Vieweg, Braunschweig.

Zeh, H.-D. (1970): On the interpretation of measurement in quantum theory. *Foundations of Physics*, **1**, 69–76.

Zeh, H.-D. (1973): Toward a quantum theory of observation. *Foundations of Physics* **3**, 109–116.

Zurek, W.H. (2003): Decoherence, einselection, and the quantum origins of the classical. *Reviews of Modern Physics* **75**, 715–775.

# Primas, Emergence, and Worlds

William Seager

**Abstract** Hans Primas was first and foremost an esteemed scientist at the forefront of quantum chemistry. But he also had abiding and deep philosophical interests, both in the philosophy of science and speculative metaphysics. This paper discusses Primas' philosophical views about the nature of emergence and ultimately the relation between mind and matter. His account of emergence has a deceptively natural link to the so-called many worlds interpretation of quantum mechanics. This link is explored and exposed as inadequate to Primas' thought. Some more speculative remarks about the metaphysics of the mind- matter relation then conclude the paper.

## 1 The Mechanistic Dream

Since the very beginnings of human thought it has been noticed that the world is made of more or less complicated things which have smaller parts (which themselves have yet smaller parts) and that the properties of the wholes depend on the properties, arrangement and interactions of the parts. This pervasive if at first doubtlessly inchoate line of thought began to be codified and made more precise just as soon as humans began to develop the intellectual apparatus required for theoretical engagement with the world.

Doctrines of atomism go back thousands of years in both Western and Eastern traditions, most especially in ancient Greece and India (see e.g. Gregory 1931; Pyle 1997; Gangopadhyaya 1981) Of course, ancient thinkers did not have a very well worked out idea of mechanism, perhaps because they lacked the rich set of technological examples, such as the pendulum clock, which enriched the thinking of the early scientists of the 17th century (see Berryman 2009).[1] But the ancients certainly advanced the common sense observation of how appropriately arranged parts

---

[1]The mechanical ingenuity of the ancients should not be underestimated however, as the discovery and eventual decoding of the Antikythera illustrates (see Freeth et al. 2006).

W. Seager (✉)
Department of Philosophy, University of Toronto, Scarborough, ON, Canada
e-mail: seager@utsc.utoronto.ca

© Springer International Publishing Switzerland 2016
H. Atmanspacher and U. Müller-Herold (eds.), *From Chemistry to Consciousness*, DOI 10.1007/978-3-319-43573-2_5

generate more complex structures and behaviors to a new theoretical and at least quasi-scientific viewpoint.

The development of modern science allowed for a more precise statement of the mechanical world view in terms of mathematical laws governing the interaction of material objects (e.g. particles). For example, the law of conservation of energy permitted the strict deduction of the outcome of particle collisions, given their initial velocities. It began to seem that nature might be nothing more than a gigantic, and gigantically complicated, pinball machine, an idea that was famously expressed by Pierre Laplace (1825/2012):

> An intelligence that, at a given instant, could comprehend all the forces by which nature is animated and the respective situation of the beings that make it up, if moreover it were vast enough to submit these data to analysis, would encompass in the same formula the movements of the greatest bodies of the universe and those of the lightest atoms. For such an intelligence nothing would be uncertain, and the future, like the past, would be open to its eyes.

This quotation is usually presented in a discussion of determinism but here the important point is the implicit idea that the world can be resolved into the "lightest atoms" and completely understood in terms of their interactions as determined by "all the forces that animate nature".

In its purest form, mechanism would endorse only a set of atomic[2] particles which interact solely by elastic collisions. An extremely precise and austere formulation of the mechanistic ideal was presented much later by C.D. Broad. He writes that (Broad 1925, pp. 44–45)

> ...the essence of Pure Mechanism is:
>
> (a) a single kind of stuff, all of whose parts are exactly alike except for differences of position and motion;
>
> (b) a single fundamental kind of change, viz, change of position. ...
>
> (c) a single elementary causal law, according to which particles influence each other by pairs...
>
> (d) a single and simple principle of composition, according to which the behavior of any aggregate of particles, or the influence of any one aggregate on any other, follows in a uniform way from the mutual influences of the constituent particles taken by pairs.

Despite its evident simplicity, notice that Broad's characterization sneaks in some features that might be regarded as suspiciously extra-mechanical. As opposed to the general scheme of an elementary causal law, isn't the only allowable interaction elastic collision between the putatively ultimate and fundamental tiny atoms of matter? But it is extremely difficult to make such a super austere scheme work. Perhaps Descartes's vortex based physics comes close but it was demonstrated quite early

---

[2]By the term "atomic" it might be understood either an absolutely smallest piece of matter or a merely contingently unbreakable and very tiny piece of matter. Most thinkers of the early modern period would have opted for the second conception if they wished to endorse atomism, since they regarded an extended piece of matter as in principle divisible, say, at least, by God.

on that systems of vortices could not generate the elliptical orbits of the planets.[3] Broad's principle of composition also suggests some constraints beyond that of the impenetrability of matter.

The additional element to the mechanical picture was that of *forces*: the general power to instill motion into matter. As is well known, even Newton regarded forces with misgivings, most especially ones that, like his own gravitational force, acted over a distance and instantaneously.[4] But Newton recognized the significance of adding forces to nature and hoped for a force based chemistry (Newton 1687/1999, pp. 382–383):

> For many things lead me to have a suspicion that all phenomena may depend on certain forces by which the particles of bodies, by causes not yet known, either are impelled toward one another and cohere in regular figures, or are repelled from one another and recede.

Every new force represents a step away from pure mechanism. Imbuing matter with mysterious powers does not accord with the goal of showing how complex structures appear simply as the result of simple units interacting according to an intelligible scheme of interaction.

Modern science has gone very far down the road of adding forces whenever convenient for explanation and with the acceptance of field theory by the late 19th century abandoned even the pretense of requiring a mechanical explanation for all effects. The pioneers of the scientific revolution would likely have recoiled from the proliferation of 'immaterial' fields and forces found in modern physics and "the forces … of contemporary microphysics would likely not have been regarded as matter by the architects of the mechanical philosophy" (Normore 2007, p. 117).

Taking a very broad and distant view of things, we can see the history of science as a grand project, which we might call the *Parts Project*. Newton famously expressed the project in terms of the correlative activities of analysis and synthesis (Newton 1730/1979, Query 31):

> By this way of Analysis we may proceed from Compounds to Ingredients, and from Motions to the Forces producing them … And the Synthesis consists in assuming the Causes discover'd, and established as Principles, and by them explaining the Phaenomena proceeding from them, and proving the Explanations.

The goal of the project was to begin with the commonsense vision of the way complex objects are constructed out of simple parts, whose arrangement and interactions explain the resulting properties and dispositions of complex objects. Commonsense observes that the world manifestly has a part-whole structure to it. The Parts Project was to show that *everything* fits into this general schema. Pure mechanism was

---

[3] Both Leibniz and Jacob Bernoulli, among others, attempted a quantitative explanation of Kepler's laws in terms of vortex theories, but neither account was fully worked out or, as was eventually realized, could be worked out (see Aiton 1972 for details).

[4] Newton acidly observed that taking his own account of gravity as revealing a property "innate, inherent and essential to Matter" which could generate instantaneous effects at a distance would be to embrace such an absurdity that "I believe no Man who has in philosophical Matters a competent Faculty of thinking can ever fall into it" (see Newton 2004, p. 102).

the first serious effort of the project. Its purity was exemplary. Its adherence to the commonsense view admirable. But its ability to actually explain the complex structures and processes of the world was woefully inadequate.

Newton conjectured a minimal retreat. Take the world as composed of material parts, atoms or atom-like units of matter, and add to them primitive powers or forces in order to explain mechanistically inexplicable interactions. Gravity is only one example and one that Newton himself was suspicious of insofar as it strayed from pure mechanism.

## 2   The Great Irony

The Parts Project inaugurated the most successful intellectual project every under-taken by the human race: empirical science in general and in particular mathematical physics. Once freed of the constraints of pure mechanism, the project raced ahead. In the mid-19th century, James Clerk Maxwell added fields to our physical ontology. Fields as such do not operate by mechanical contact, though Maxwell initially made considerable efforts to devise mechanical models of the electromagnetic field.[5] There was much worry that without such models a vicious gap in intelligibility would ensue but over time such scruples faded away. At least one could content oneself that the electromagnetic field was generated by material sources of charge even if it it did then embody its own causal powers. The Parts Project remained viable as the 19th century drew to a close. A number of prominent physicists went so far as to declare that the scientific metaphysics of the (albeit extended) mechanical world view was virtually complete (see e.g. Badash 1972; Schaffer 2000).

But the Parts Project soon thereafter collapsed. It was exploded by the development of quantum mechanics. The world does not resolve itself into elementary, independent objects which fit together under simple laws of interaction. The most successful intellectual project ever undertaken by the human race actually ends with the collapse of the project's initial motivating idea. This great irony was emphasized throughout his philosophical writings by Hans Primas, from a number of different viewpoints. For example (Primas 2007, p. 8):

> Modern quantum mechanics put an end to atomism and hence to reductionism: The so-called "elementary particles" (such as electrons, quarks, or gluons) are patterns of reality, not building blocks of reality. They are not primary, but arise as secondary manifestations, for example as field excitations, in the same sense as solitons are localized excitations of water, and not building blocks of water.

Much earlier Primas (1998, p. 88) wrote:

---

[5]For discussion of various aspects of Maxwell molecular vortex model see Siegel (2003), Chalmers (2001), Dyson (2007). It seems that Maxwell at first regarded these with, as Siegel puts it, "onto-logical intent" (Siegel 2003, p. 56) but came to see them later as heuristic aids to understanding. For Maxwell's own presentation of his model see Maxwell (1890/1965, pp. 451ff).

The historical idea that the material world is already structured by some kind of interacting "atoms" is in sharp contradiction to basic insights suggested by quantum mechanics.
According to quantum theory the material world is a whole, a whole which is not made out
of independently existing parts. As a rule, separated subsystems of a quantum system do not
exist.

It remains very difficult to grasp fully the implications of these ideas which replace
rather than modify the mechanistic account of the world, even in its extended form.
Most philosophers, scientists and even physicists struggle to come to grips with the
idea that the world in not constructed from fundamental micro-objects. The flood of
popular modern physics books does little to dispel the idea that the world is made
of small, discrete and independent objects, and Primas conceded that "in spite of
the fact that quantum mechanics put an end to atomism, modern science is still to a
large extent based on an atomistic ontology" (Primas 2007, p. 8). Even though most
physicists would probably agree with David Wallace's acidic assessment that "the
popular impression of particle physics as about the behavior of lots of little point
particles whizzing about bears about as much relation to real particle physics as
the earth/air/fire/water theory of matter bears to the Periodic Table" (Wallace 2013,
p. 222) there remains a widespread impression that the world is made out of tiny
objects which physics tells us about.

# 3 Emergence

There is, of course, a large assumption that underwrites the fatal diagnosis of the Parts
Project which is that quantum mechanics (QM) is true or at least "true enough" that
its non-mechanistic and holistic picture of the world will be sustained in successor
theories. It is impossible for anyone to say with absolute certainty that QM will form
the core of all future science or that it will not be entirely eclipsed in some huge
scientific revolution. But it would take someone very brave to bet against QM.

QM is the most thoroughly scientific theory of all time, by a wide margin. Recently,
some of these tests have taken a remarkable form. It is a curious fact that the features of QM that are most deeply antithetical to the mechanistic view of the world
are accessible to experimental investigation. This is sometimes called experimental metaphysics, and it got itself onto a firm footing after the work of John Bell
(see e.g. Bell 1987, Chaps. 1 and 2). Through a somewhat intricate but conceptually straightforward proof, Bell showed that no mechanistic account of nature could
duplicate the predictions of QM. The crucial aspect of mechanism here is that of
local interaction between independent units or "hidden variables" which are supposed to underlie the empirical regularities explained and predicted by QM.[6] This
discrepancy in the predictions of local realistic theories and QM can be and by now

---

[6]The idea that the world is made of particulate units is not refuted by Bell's result, if the units lose
their independence and are, so to speak, in a kind of universal communication with one another.
Theories such as this go back to the early days of quantum mechanics with the "pilot wave" of Louis
de Broglie in the 1920s. Since David Bohm's (1952) rediscovery of the de Broglie approach it has

has been extensively tested, with results uniformly and completely in favor of QM (for some recent results see Hensen et al. 2015, Poh et al. 2015; Wikipedia has a nice history of the relevant experiments[7]).

But another peculiarity of QM is that even if we grant, on the grounds of its vast empirical success, that it presents a reasonably accurate account of reality, it remains unclear exactly what kind of reality it is portraying. This is the problem of *interpreting* QM, a problem with little or no counterpart in any other part of science. How could a mature theory used by thousands of scientists every day be so interpretively opaque?

The best guess is that QM strains our ability to conceptualize an ontological scheme which incorporates all of QM's bizarre theoretical features. This has led to a host of interpretations which run the gamut from micro anti-realism to many worlds ultra-realism.

Micro-antirealism is the view that QM does not describe and is not intended to describe an existing microscale world at all. Rather, what exists is the macroscopic domain of manifest experience which is amenable to description in classical terms. QM then provides us with rules for predicting the evolution of features in the manifest realm, or perhaps can be regarded as encoding the intrinsically probabilistic epistemic limitations observers confront when attempting to make such predictions (roughly speaking, the former view is more like Bohr's so-called Copenhagen interpretation while the latter, very closely related, has been labeled quantum Bayesian interpretation[8]).

Bohr's perceived micro anti-realism was once a kind of orthodoxy but has fallen into disfavor more recently amongst philosophers of science and physicists interested in quantum foundations. A particularly stark description has been given by Tim Maudlin (2010, p. 127):

> Bohr sometimes sounds like this: there is a classical world, a world of laboratory equipment and middle-sized dry goods, but it is not composed of atoms or electrons or anything at all. All the mathematical machinery that seems to be about atoms and electrons is just part of an... apparatus designed to predict correlations among the behaviors of the classical objects.

While it is far from clear that this is a completely fair characterization of Bohr it can stand as a characterization of micro anti-realism, and it is anathema to most current philosophers of science. Maudlin's own blunt assessment is simply that "I take it that no one pretends anymore to understand this sort of gobbledegook ..." (Maudlin 2010, pp. 127–128). It is interesting that at least to a certain extent, and long before Maudlin wrote, Primas took a similarly stark view of Bohr's view of the micro-world, by contrasting it the viewpoint of practicing chemists: "Chemists never have adopted Bohr's view that microphysical objects do not exist" (Primas 1981/2013, p. 158).

---

(Footnote 6 continued)

seen extensive development; see Holland (1993) for technical details, Bohm and Hiley (1993) for a more general overview and some philosophical extrapolations). The point is that the de Broglie-Bohm approach does not reinstate the mechanistic dream.

[7]See https://en.wikipedia.org/wiki/Bell_test_experiments.

[8]Bohr's philosophy of science is difficult to spell out precisely but see Murdoch (1989). Quantum Bayesianism was developed over a number of publications by Carlton Caves, Christopher Fuchs and Rüdiger Schack; for an overview see Timpson (2008).

I am not, myself, so sure that Bohr should be relegated to the dustbin. After all, the route to the micro-world begins with our everyday observation that common physical objects are made of parts, which have further parts, etc. A brick wall is made of bricks, and the bricks themselves are made of grains of sand, and the grains of sand are made of ... But we have already seen that this is the pathway that leads to the Parts Project, and we know how that turned out.

Whatever we think of the micro-world, one core lesson of QM is that it is not anything at all like a world of small objects zipping about and independently interacting to compose more complex entities in anything like the way grains of sand compose bricks. There must indeed be some link from the story which QM tells to our familiar world of manifest experience. But this link from whatever the quantum realm is to the classical or manifest world of experience cannot be the dreamt of system of whole to part decomposition because, in Primas' own words, "according to quantum mechanics, the material world is a whole, *a whole which is not made out of parts*" (Primas 1995, p. 611, original emphasis).

The linkage from how QM describes its part of the world (micro-world or not) to the world of manifest experience is the general problem of *emergence*: how to construct or retrieve the world as we experience it from the peculiar world QM presents us with. The problem of emergence is ancient because of the common observations that lead to the Parts Project. It is evident, for example, both that birds are not made out of more birds and that birds are made out of parts. So the question naturally arises how the non-bird parts "combine" or come together to produce a bird. The ancient pre-Socratic philosophers struggled with this and came up with the basic dichotomy: inherence versus origination (see Mourelatos 1986). Advocates of inherence cleave to the dictum *ex nihilo nihil fit*; whatever emerges must in some substantial sense already be present in the submergent base. Defenders of origination hold that at least sometimes emergent features are genuine ontological novelties which are not determined by the state and laws governing just the submergent features.

Much, much later—in the late 19th and early 20th centuries—arose a sophisticated account of emergence which opted for origination. Since most of the thinkers associated with this view were British, it has come to be known as British Emergentism (for an overview see McLaughlin 1992). The British Emergentists were realists about the physical world and held that everything was determined by the fundamental physical features of the world. But they also held that some features were merely the causal result of certain configurations of matter, where the causal laws which related the submergent to the emergent were themselves fundamental. The emergent features were not determined by the laws governing just the basic physical features. Instead, the laws of emergence were "free additions" to the world, or what C.D. Broad called "trans-ordinal laws" (see Broad 1925, pp. 77ff) and what John Stuart Mill had earlier labeled "heteropathic" effects (see Mill 1843/1963, pp. 443ff)

Philosophers like to use a theological metaphor here. What did God have to create in order to create the world? If one follows inherence about emergence then the answer is that God simply needed to create the laws of fundamental physics and arrange the fundamental physical features in some suitable initial condition. Everything else (stars, planets, geology, life, mind) would follow, strictly determined by the ongoing purely physical development of the world after its creation. On the other hand, one who takes the origination line on emergence would hold that God was not finished His creative work simply in virtue of His initial laying down of the fundamental physical laws and features. In addition, God would have to institute certain "laws of emergence" (Broad's inter-ordinal laws) which would come into effect whenever physical configurations arose of the proper complexity and which would originate some genuinely novel feature. One might also put the point in terms of whether all laws of nature stem from the laws of physics alone (plus, perhaps, the arrangement of physical features if, as it may, be some laws are contingent upon matter being arranged in the appropriate way).

The heyday of British emergentism was the early 20th century, up to about 1925. They regarded their origination based account of emergence as almost obviously true and their lynchpin, supposedly uncontroversial example was chemistry. Here is Broad (1925, pp. 62–63):

> We will now pass to the case of chemical composition. Oxygen has certain properties and Hydrogen has certain other properties. They combine to form water, and the proportions in which they do this are fixed. Nothing that we know about Oxygen by itself or in its combinations with anything but Hydrogen would give us the least reason to suppose that it would combine with Hydrogen at all. Nothing that we know about Hydrogen by itself or in its combinations with anything but Oxygen would give us the least reason to expect that it would combine with Oxygen at all. And most of the chemical and physical properties of water have no known connexion, either quantitative or qualitative, with those of Oxygen and Hydrogen.

Rather unfortunately for Broad and the rest of the British emergentists, 1925 was the year that QM was put on a secure theoretical footing and it began to be clear that the fundamental physical features that make up oxygen and hydrogen actually do determine that they will combine in a ratio of 1-to-2 and that the qualitative features we observe of water are similarly determined by the underlying physical constituents.

By 1929, Paul Dirac could seriously proclaim that (Dirac 1929, p. 714):

> The underlying physical laws necessary for the mathematical theory of a large part of physics and the whole of chemistry are thus completely known, and the difficulty is only that the exact application of these laws leads to equations much too complicated to be soluble.

Now, the correct characterization of the relation between physics and chemistry remains controversial. Primas had much to say about this, more than I have space or the ability to go into. Primas certainly denied that chemistry could be *reduced* to physics, in the distinctively philosophical sense of reduction as developed by Ernest Nagel (see e.g. Nagel 1961) and others. This formal conception of reduction envisions a translation scheme according to which the reduced theory (here, chemistry) could be completely rewritten in terms of the reducing theory (here, physics) Primas regarded such philosophical accounts of reduction as insufficiently well defined to be of use in real scientific work (Primas 1998, p. 83).

However, it does seem clear that Primas did not endorse the kind of radical onto-logical origination espoused by the British emergentists. The physical world does have a fundamental structure which determines everything else, but the relations between theories is highly complex and dependent on creative abstractions, mathematical procedures, approximation techniques, experimental selection and other acts of mind: a host of factors which Primas included in the general notion of *contextuality*. Emergence can then be characterized thus (Primas 1998, p. 83):

> Emergent properties are not manifest on the level of the basic theory, but they can be derived rigorously by imposing new, contextually selected topologies upon context-independent first principles.

Two central concepts developed by Primas to explain the quantum to classical transition are those of "endophysics" and "exophysics". The context independent domain is that of endophysics; the domain of contextuality is exophysics (see Primas 1994).[9] Exophysics is derivable from endophysics, once the context has been fixed.

If all possible contexts of experimentation were mutually compatible then exophysics would be reducible to endophysics. Emergence would then simply be a reflection of complexity and our own epistemic limitations. One of the astonishing lessons of QM, however, is that it is impossible, even in principle, to perform measurements simultaneously on all observable or measurable properties of physical systems. In terms of the distinction between endophysics and exophysics, this means that there is no standpoint from which all exophysical features can be derived purely from the endophysics, even though it is true to say that the endophysical realm is what is ultimately real and fundamental. QM forces us to recognize that even though "the first principles of physics are intended to give ... a context-independent description of the material world" (Primas 1998, p. 85) this will not yield access to the world we directly experience. To move from the "intrinsic description" of the world as described by the first principles, which "makes no reference to other physical systems" (how could it?), we have to impose a context, for example, of measurement.

The world of exophysics is like a set of tiles that cannot be laid down together to cover the floor, even though each tile does cover some part of the floor and no part of the floor is not covered by some tile or other. We need contextualization to select, so to speak, one of the tiles to lay down. One of the most remarkable aspects of Primas' view was the way he linked contextualization to perspectives and pattern recognition. The exophysical world is a set of patterns which can only be discovered from particular perspectives. Going back to the 1981 first edition of Primas (1981/2013), the emphasis on patterns anticipates the work of Daniel Dennett and the subsequent development in the philosophy of science of the so-called ontological structuralists (see Dennett 1991; Ladyman et al. 2007). Dennett's conception of patterns is entirely classical and indeed mechanistic at heart (his main example is John Conway's[10] "game of life" cellular automaton). Primas' system of patterns inherits the non-classical nature of QM. Patterns are recognizable regularities that arise in experimental (or observational) contexts. While these contexts are themselves

---

[9]Primas' conceptions of endophysics and exophysics are developed from the initial formulation of David Finkelstein (1995). Interesting philosophical discussions of Primas' notion of endo- and exophysics can be found in Shimony (1999) and d'Espagnat (1999).

[10]Conway's "game of life" was first introduced widely to the world by Martin Gardner (1970).

classical domains, there is no way to arrive at a description of the total system by "summing" or "combining" the set of contexts—they are incommensurable.

The core metaphysical vision of the world at work is that of an underlying monistic and holistic reality, perhaps reminiscent of what Spinoza called God. This endophysical fundamental reality is not manifest in experience. It is entirely independent of mind and is fully objective ("endophysics refers to a subject-independent reality"). Our most fundamental theories strive to describe the endophysical reality in terms of "metaphysical universal laws", but the "endoentities ... are hidden from us and ... not directly observable". The realm of the observable is that of exophysics which "aims to give us empirically adequate descriptions" (all the foregoing quotes from Primas 1994, p. 168).

The world of manifest reality is the exophysical world. It is, to a first approximation, a classical world that appears atomistic and mechanistic. Primas was able to express the essence of classicality by a distinction between systems describable in terms of Boolean logical structures (classical) versus those which could not be so described (quantum). The field of quantum logic has long recognized that there is no way to encode quantum theory in a Boolean logic but Primas emphasized the way that the manifest reality of experience, the realm in which experimentation takes place, must be describable in Boolean terms but that it is impossible to combine these Boolean descriptions into one overarching description of the entirety of reality (see e.g. Primas 2003, 2007; Atmanspacher and Primas 2003).

The great irony discussed above is the dissolution of the project which attempted to take the exophysical for the endophysical. There is no way to render the endophysical totality in an exophysical picture. Classical (or semi-classical) domains are exophysical features emergent from the underlying holistic endophysical reality where this emergence is conditioned by the perspective of the experimenter via choice of apparatus and context, revealing a pattern. However, the system of all such patterns is not coherent; the world cannot be regarded as the sum of patterns into an overarching world in which they all appear.

In recent philosophy of QM, there is a view which one could be forgiven for identifying with Primas' account. It bears many affinities with our sketch of the relation between the endophysical and the exophysical. Yet, Primas did not accept this view even though he had been originally attracted by it.

# 4   Many Worlds

This view which superficially resembles Primas' goes by several names: the relative state interpretation or QM, the Everettian interpretation and the many-worlds interpretation. It was invented by Hugh Everett (1957). The core idea is that we ought simply to accept what the mathematics of QM seems to be telling us. This mathematics holds that there is never any sudden and discontinuous transition of the quantum mechanical wave function which makes one of its components "become real". As is very well known, quantum systems are generally in states which are described by superpositions of states which represent observable properties having definite values. For example, an electron might be in a state which is the superposition of two possible spin states. No electron has ever been directly observed to be in such a state. Whenever measured, an electron reveals itself to be in a quite definite spin state. The orthodox explanation for this peculiar state of affairs is that, upon measurement, the state of the system transitions, or *collapses*, into one of the components of the superposition. Orthodoxy is enshrined as an "axiom" of QM called the projection postulate.[11]

Everett's theory eliminates the projection postulate. The quantum wave function always and everywhere evolves according to the deterministic mathematics which is the core of QM, as in the Schrödinger equation. How then to explain the failure to ever observe a system actually in a superposition? Everett took the bold step of accepting that the observing equipment and the experimenter as well would evolve into a superposition no less than any other physical system.

If we take the somewhat audacious view that the entire universe is a physical system then, cosmologically speaking, there is a "universal wave function" which, so to speak, evolves into an immense superposition which is a foliation of all possible state evolutions. The branches of the foliation include the system being measured, the measuring instrument, the human experimenter and indeed the entire environment which ever has interacted with any component of these components–in short, the entire universe we inhabit is but one component of a vast all encompassing superposition of all physically possible evolutions. Defenders of the many-worlds interpretation of QM like to say that it is not really an "interpretation" since it is simply what the mathematics tells us. The metaphysical structure of many-worlds is just "read off" the mathematics.

Although Everett's many-worlds interpretation was for a long time a decidedly minority position amongst both physicists and philosophers of physics, it has enjoyed a remarkable renaissance in the 21st century. Most especially, there has grown up the so-called Oxford program in which a number of philosophers of science, mostly indeed housed at Oxford University, have produced an impressive defense of the many-worlds interpretation (see e.g. Wallace 2012 and its references). The Oxford program has addressed directly what many regard as the most important objections to the many-worlds interpretation.

There are two fundamental challenges facing the many-worlds interpretation which were noted almost as soon as Everett announced it (in fact, Everett recognized them in his seminal work and attempted responses). The first is the Probability Problem. QM has an algorithm for determining the probability of any observation which is called the Born Rule (formulated by Max Born 1926). In the simplest and historically significant case, the rule states that the probability of finding a particle at a certain position is the square of the amplitude of the wave function at that point in

---

[11] John von Neumann (1932/1955) articulated and attempted to justify the postulate in his magisterial monograph. It has been the subject of a vast literature which has been largely negative because of the unattractive way that the postulate simply asserts that there will be a sudden break with the otherwise smoothly predictable evolution of a quantum system when a hard to define event of "measurement" occurs.

space. For the case of a particle in a superposition of two spin states the probability of observing the particle in a particular spin state is the square of the "weight" of that component. For example, such a state might be written as

$$\sqrt{\frac{3}{4}}A^+ + \sqrt{\frac{1}{4}}A^-$$

In this case the Born Rule tells us that there is a $\frac{3}{4}$ chance of finding the system in state $A^+$ and a $\frac{1}{4}$ chance of finding it in the $A^-$ state. The Probability Problem is now evident. If the world 'splits' upon a measurement there are only two possible outcomes and the many-worlds interpretation holds that both actually occur (along with a similar dual splitting of everything connected to the system under observation, most notably the experimenters themselves). If both outcomes occur, how can there be any differentiation in the probability of the outcomes?

To put the point starkly, what, according to the many-worlds interpretation of QM, is the difference between the above state and

$$\frac{1}{\sqrt{2}}(A^+ + A^-)\,?$$

The amplitudes seem to be metaphysically otiose. They make no difference to the way the world actually evolves.

Before discussing the Probability Problem further, let us turn to the second problem afflicting the many-worlds interpretation. This is a problem of emergence. Although the range of quantum possibilities is truly vast, we only ever seem to observe a world that is to a good approximation classical. Objects have highly definite positions and never just disappear and reappear in another location, objects do not migrate through walls unscathed, etc. How is the deep strangeness of the quantum world suppressed or eliminated in the world(s) that we experience?

This is a problem that Everett himself tackled and pointed the way towards a solution. Since then huge amounts of work have been done addressing the question of how classical branches appear and dominate the foliating superposition of all possible states which the many-worlds interpretation asserts is the true reality of things. The key concept is that of decoherence: the general tendency for quantum superpositions to lose their internal correlations as they interact with the environment. This is most evident in the case where the "environment" is a measuring device.

The famous two slit experiment is a perfect illustration. This experiment is so well known that it hardly needs describing but I will recall its structure here very briefly. Imagine a beam of particles directed at a screen on which there are two very small openings through which they can pass. Beyond this barrier screen lies a detector screen on which we can observe where the particles impact. QM predicts that the pattern of impacts will not be the simple addition of impacts from passage through each slit (a kind of "two hump" distribution that would be the result of shooting classical particles, such as bullets from a machine gun) but will rather exhibit a system of bands of impacts. This is caused by the quantum interference effects of the two possible paths. However, if a detector is placed so that we can determine which slit a particle passes through, then the band pattern disappears to be replaced with the two hump distribution. Mathematically, the two detector states are orthogonal: any terms in an equation where they combine will go to zero. Once we put the detectors into our two-slit experiment, the interference terms will contain combinations of the two detector states and these terms will disappear. And so the interference has been eliminated, as we can observe … or has it?

A complete quantum description of the experimental setup with detectors would predict that the (experiment + detector) system would itself go into a superposition. If we could somehow, and it would already be very difficult, arrange the appropriate experiment on the combined (experiment + detector) system it too would exhibit interference effects. In order to actually do this, we would have to completely isolate the (experiment + detector) system to preserve its quantum coherence. This is very difficult to do and the more complex the system and the longer the time period of observation the greater the degree of interaction between the (experiment + detector) system and various parts of the general environment. In effect, under normal conditions the environment is acting something like a detector, watching over, so to speak, its own parts. It can be shown that most environmental states will be effectively orthogonal to each other and they will enforce the loss of quantum coherence. Distinct quantum effects will thus tend to be suppressed.

Of course, many of the very most complex and highly relevant parts of the environment are the brains of observing scientists. These 'physical devices' will themselves be in thorough interaction with huge number of environmental parameters, so we would expect that the observation of quantum effects by human observers will also be suppressed. Brains too will tend to act classically.[12] This is the general scheme of decoherence, the details of which are involved, intricate and have been developed with great precision and sophistication (a seminal sourcebook is Joos et al. (2003); see Wallace (2012) for philosophical discussion).

From the perspective of the many-worlds interpretation of QM, decoherence strongly suggests that almost all[13] the branches in the universal foliation, and certainly almost all of them with physically complicated conscious observers, will appear to be a classical world with definite objects having determinate positions and velocities.

It now seems that the decoherence approach will eventually provide a full understanding of how classicality emerges from the universal wave function postulated by the many-worlds interpretation of QM. What is interesting here is that there is an almost irresistible mapping from the decoherence approach to Primas' own account of the emergence of classical systems. The equation is simply this: endophysics = the universal wave function, exophysics = the elements of the foliation, or the branches,

---

[12]While the brain must at bottom be a quantum system (since *everything* is), it remains very controversial whether distinctive quantum effects are a significant component of brain function. See Hameroff and Penrose (1996) for a positive view; Tegmark (2000) and Eliasmith (2000) for criticism.

[13]I am using the phrase "almost all" colloquially but it may also be true in the mathematical sense that the elements of the universal foliation, which are an uncountable infinity, are all save for a set of measure zero essentially classical. I don't know whether this is provable.

or the "worlds" of the many-worlds interpretation. The points of similarity between this interpretation and Primas' views are manifold. The branches are individually classical (or virtually classical), as are the exophysical systems. The branches cannot be combined or summed into one coherent world which is either available to or manifest in ordinary experience and yet the totality of them is the underlying and fundamental reality of things, in line with Primas' account of the endophysical. Following from the basic structure of QM, the branches are contextual and perspective based for the particular observables that will take a determinate value in each branch depend on the measurement setup plus environment, choice of experimenter, etc. In this way again they are very similar to Primas' exophysical systems.

If we step back and regard the universal wave function, the vast superposition of all possible states, we see more links to Primas' views. The totality of the universal wave function is decidedly non-classical exhibiting a holistic character with deep entanglement throughout (albeit the correlations between components are "smeared out" into the wide environment of each branch). Metaphysical dependence runs from the whole to the parts rather than the reverse, in a way reminiscent of Primas' endophysical world (and Spinoza's holistic monism).

In light of the at least interesting correspondence between the many-worlds interpretation and Primas' endo-exo division, it would be worth exploring Primas' attitude towards the many-worlds interpretation. This project can begin on an optimistic note. In his early philosophical writings, Primas took a positive view of Everett's theory. In 1981 he wrote that that Everett's account was "superior in logical economy" and, more significantly, that it provides "a more intelligible pattern of the world" (Primas 1981/2013, p. 135).

I have been reliably led to believe that over time Primas' positive attitude towards the many-worlds interpretation soured, but I have not been able to find anywhere in Primas' writings where he engages in any sustained criticism of Everett's views (and I would appreciate any tips about where to look). Primas early on noted that "the conclusions of the Everett interpretation may be considered as bizarre" (Primas 1981/2013, p. 135) but that would hardly, and especially for a thinker like Primas, count as a cogent argument against it (he immediately adds to the last quote: "novelty and repugnance are not valid arguments").

It may be that Primas was unhappy with how Everett's original views were transmuted, mostly at the hands of Bryce DeWitt, into a theory that explicitly endorsed the picture of there literally being many worlds. Jeffrey Barrett has noted that "there is no mention of splitting worlds or parallel universes in any of Everett's published work" (Barrett 2011, p. 277). Everett's own view seemed more akin to a monistic holism in which all possibilities (consistent with the wave function which specifies the quantum nature of reality) are real. Within such a picture, Everett strove to show that individual observers, or individual consciousnesses, would experience a world in which particular measurements would have definite outcomes—there would be no *appearance* of superposition states despite the fact that every observer is a part of the vast superposition which is reality. As we shall see below, this monistic viewpoint has affinities with Primas's own metaphysical outlook, but it is hard to associate it directly with Everett himself.

For at the same time, it is difficult to see that Everett's view is all that different from what has become the standard many worlds version of it. For example, Everett describes his view in terms of "splitting observers" (see Barrett 2011, p. 288). It seems pretty clear that if observers can split they are going to take their worlds with them. Indeed, in the transcript of Everett's presentation at the 1962 Xavier conference on the foundations of quantum mechanics we find Everett saying (Barrett 2011, p. 292):

> ...it's a consequence of the superposition principle that each separate element of the super-position will obey the same laws independent of the presence or absence of one another. ...Each individual branch looks like a perfectly respectable world where definite things have happened.

Barrett claims that Everett did not seem to particularly care how this aspect of his view was described for the interesting reason that Everrett had a distinctly empiricist and somewhat anti-realist outlook on the philosophy of science. Everrett would have denied that any scientific theory could claim to be *the* true view of reality. Instead, Everett writes that, "any physical theory is essentially only a model for the world of experience, we must renounce all hope of finding anything like 'the correct theory'" (Everett 1973, p. 134). Thus it is somewhat forced, and certainly not mandated, to equate Everett's view with a Spinozistic kind of monism (with the universal wave function replacing Spinoza's "God"). So it remains somewhat puzzling what feature of Everett's account precipitated Primas's eventually distancing himself from the theory, given his initial positive reaction.

We may find the beginning of a possible solution to this puzzle if we go back to the first of the two major challenges facing the many-worlds interpretation: the Probability Problem. Recall that this is the difficulty of justifying Born's rule of assigning probabilities to outcomes of quantum measurements. One might have thought that one of, if not the, core idea in our conception of probability is the distinction between the possible and the actual. Because many things are possible but only one can be actual, it is natural to seek some way to gauge the chances of any specified possibility becoming actual. The "gauge of chances" is just what we call probability. But in the many-worlds picture of reality there is no distinction between the possible and actual: if something is a possibility it will be an actuality. There are no merely possible, unactualized branches in the foliation of the universal wave function. This is a deeply counterintuitive conception of reality, very much at odds with how we organize our own experience.

So there has naturally been a great deal of effort expended on showing that the Born rule can be vindicated. Early efforts go all the way back to Everett's own work. It can be shown with considerable rigor that branches that violate the Born rule will be branches with low quantum amplitude. But, as noted above, without a connection already established to probability, amplitudes are just numbers assigned to, in this case, branches in the foliation of the universal wave function, by an arcane mathematical procedure. Why should we care about them?

The Oxford Program takes this critique to heart and jettisons discussion of objective probability. If the distinction between possibility and actuality is empty, we could recast probability in terms of subjective degrees of belief (a key concept in a venerable research program in any case). Beginning with work of David Deutsch (1999) and further developed by David Wallace (2007), a remarkable link between the Born Rule and decision theoretic considerations has been forged. Basically, Deutsch and Wallace aim to show that rational agents should assign their degrees of belief according to the Born Rule. This is not the place to delve into the burgeoning literature on this issue, but broadly speaking, the proofs offered by Deutsch and Wallace either ignore or deny the claim that adding the genuineness or reality of all the branches should affect our predictions and preferences.[14] But it seems to me that if we are to regard all the branches as equally real, we have to or should take them into account in our decision making.

If all this seems very abstract, let me give a simple, but fanciful, illustration of how quantum amplitudes could intelligibly operate in our decisions in such a way as to justify the Born Rule. Imagine, if you can, that human personal identity is a fundamental metaphysical feature of the world, no less than anything else you regard as fully objective.[15] So how should you then think of your own future in the branching structure postulated in the many-worlds interpretation? Each branching will lead to many copies of yourself but by hypothesis only one of these will truly be you. What if the quantum amplitudes were a measure of the likelihood of you ending up in a particular branch? Then it would be obvious that you should apportion your subjective beliefs according to amplitude. If you face a measurement process with unequal weights, it will really be more likely that you, yourself, will end up in the more heavily weighted branch observing the more heavily weighted outcome.

Notice that this thought experiment reintroduces some genuine uncertainty about the future into the picture, which the standard many-worlds interpretation has eliminated. That explains why it immediately offers an intuitively attractive link between the quantum amplitudes and probability even if it is perhaps metaphysically extravagant (it also has a hidden bias of self concern built into it if you stop to think of it). – Needless to say, our scientifically minded philosophers are not attracted to the idea of an objective, presumably substantial, self. We will simply regard it as an illustrative exercise.

It is easy to think up examples where differences in the way branches are created in measurement seems to matter a great deal. A typical example is the biased quantum coin flip. Let us set up a quantum experiment with two possible outcomes (call them Heads and Tails) and set the amplitudes so that the quantum mechanically calculated probability of Heads is one in a trillion. Would you pay $10 to play this game: if Heads comes up you get $1 billion, Tails you get nothing? Orthodox reasoning and commonsense prudence are both strongly opposed to your participation in this game.

---

[14]This denial is enshrined in what Wallace calls the equivalence principle (Wallace 2007, p. 318) which asserts that all that matters to assigning subjective uncertainty about some proposition, $P$ is the quantum amplitude of $P$ irrespective of, say, the way that $P$ is observed or measured to be true. That means that the number of worlds "generated" by the measurement of $P$ is irrelevant to subjective uncertainty. This seems peculiar, since many lives, including those of our quantum descendants, will hang in the balance of how many branches are pumped out by a measurement. This approach has, of course, been criticized, notably by Albert (2015) and Kent (2010).

[15]This option is sometimes called that of "primitive identity over time" (see Greaves 2004).

But the many-worlds picture suggests otherwise. After the measurement one of you will be very rich and one of you will be out $10. Obviously you should play.[16]

There is also the bizarre problem of quantum suicide (see Moravec 1988, p. 188ff). Would you play Russian roulette for a big prize? Imagine a version of Russian roulette in which death is instantaneous and painless. If you played with a quantum gun that had two outcomes (death or life, in short) then no matter what the amplitudes of the two outcomes were, you are guaranteed to survive and live on with the big prize. We can alter the game by adding, in principle, any number of outcomes leading to death and only one leading to life (that is, the death outcome can be linked to some quantum measurement process with a huge number of possible values). Now, say, millions of my descendants die off yet we are supposed to believe that this should make no difference to how I regard these situations.

An instructive illustration of the oddity of using the Born Rule to set subjective probabilities and inform decisions can be constructed from the bizarre conceit that informs the film *The Prestige*. In the movie, a magician discovers a teleportation machine. He uses it to develop an astonishing magic trick in which he miraculously transports himself across the stage. There is one horrible drawback: the machine creates a duplicate of the teleported object. So every time he performs the trick, there is the problem of what to do with the extra duplicate. The magician thus has his newly created and unwanted duplicate drowned in a locking tank of water under a hidden trap door. At one point, the magician muses to himself that he was always terrified that he would end up in the tank of water instead of appearing across the stage in the target cabinet. This seems like a very odd remark unless one believes in a metaphysically substantial self that has to go "into" one or the other of the duplicates. More realistically, one supposes that *every time* the trick is performed the drowning duplicate should exclaim: "oh no, I'm the drowning one this time".

But now, suppose that the teleporter/duplicator is a quantum device. We can suppose it creates a not quite perfect duplicate to evade the no-cloning theorem and we can also suppose that neither created copy is a exactly like the original to evade an obvious way to track identity over time. Let's say that one duplicate, $D1$, has a new tiny mole on the left cheek and the other, $D2$, has one on the right cheek. With considerable abuse of notation we can write the desired quantum state as:

$$\alpha \left( D1_{tank} \otimes D2_{cabinet} \right) + \beta \left( D1_{cabinet} \otimes D2_{tank} \right).$$

If we make $\alpha$ large enough we can make it such that the magician should (according to the view of the many-worlds interpretation we are considering) expect that D2 will end up in the cabinet, safe and sound. But how can the magician guarantee the he will be D2? By adjusting the weights, the magician can project himself, so to speak,

---

[16]As many thinkers who have contemplated the many-worlds interpretation have pointed out, it is actually very difficult to count the number of branches that will be generated by a measurement, just because of the vast number of connections between the measuring device and the rest of the world. So it is somewhat naive to analyze this quantum game as leading to just one rich you and one very slightly poorer you. But there is no reason at all to think that the outcome won't have equal numbers of rich yous and slightly poorer yous, so the point of the analysis stands.

into the non-drowning successor with arbitrarily high probability. Modern many-worlders deny there is anything like primitive identity over time or a substantial self, but they do hold that one should use the Born Rule derived probabilities to set one's expectations and help one form decisions about actions. How, then, can the magician project himself into the cabinet? Simply by adding another quantum measurement. To make it vivid, suppose that the duplicates will be shown a colored card as they are created in either the tank or the cabinet (say either a red card or a green one). Which color they see is determined by a quantum process. The magician arranges that the probability that D2 will see the green card is extremely high no matter whether D2 ends up in the tank or the cabinet. So the magician should expect to be D2 after duplication and since $\alpha$ is very high should expect himself to be in the cabinet.

This should drive home the oddity of the claim that worlds as such don't matter. In this case, there are just as many magicians drowning as surviving. It is entirely unclear why the amplitudes should make one feel even one bit safer about engaging in this magic trick.[17]

## 5   Complementarity

I have no idea whether worries about the Probability Problem lay behind Primas' worries about the many-worlds interpretation. The philosophical core of the problem has to do with the distinction between actuality and possibility, or potentiality, and also with the question of whether the world is subject to genuine temporal becoming. And these issues did come to have importance for Primas in his later philosophical writing (e.g. Primas 2003, 2007). There we see that Primas seems to have moved towards a yet more radical view of reality that has interesting affinities for a dual-aspect picture inspired by Wolfgang Pauli and Carl Jung. Such a view avoids the claim that QM itself provides the correct metaphysical account of reality, thus relieving some of the pressure that leads to the many-worlds interpretation because there is no compelling need to regard QM as providing a complete picture of reality.

Pauli had striven towards such a dual aspect view of nature, for example in a letter to Jung writing that "physis and psyche are probably two aspects of one and the same abstract fact" (Meier 2001, p. 159). But the lesson which QM teaches is that dual aspects can stand in a very special relationship, that of complementarity: "It would be most satisfactory if physis and psyche could be conceived as complementary aspects of the same reality" (Pauli 1952/1994, p. 260). This view is explicitly endorsed by Primas when he writes that "all physical theories at our disposal are essentially incomplete theories: they are incapable to deal with the complementarity of matter and spirit" (Primas 1995 p. 611).

---

[17]It should also add to worries about the intelligibility of the primitive identity over time approach. The link between the quantum amplitudes and personal identity seems entirely arbitrarily imposed, without even a hint of any coherent connection between quantum measurement and the location of the self in the quantum world.

There are two key features of complementarity that matter here. The first is that complementary attributes are ontologically equal, neither reduces to the other. Second, complementary attributes do not reduce to underlying fundamental attributes; they are co-fundamental. Here is obviously a break with the views of most philosophers who accept the many-worlds interpretation of QM, for they regard it as a way to reduce the mental to the physical. The many-world interpretation is supposed to be part of the general advance towards a thorough physicalism, removing some of the mystical garbage (or, in Maudlin's term, gobbledegook) that has accrued around QM. Applied to the mind-matter relation, complementarity suggests that the mental and the physical are co-fundamental features (attributes) of some single underlying substance which is itself un-representable (as Pauli sometimes called it: *unanschaulich*).

It is natural to ask which aspects of the mental require a complementarity based understanding. And the natural answer is that it is consciousness or the subjective elements of experience (the "what it is like" of experience famously described by Nagel 1974). Typically, these are considered to be the qualitative features of experience, especially sensory experience but the subjective elements of consciousness are multiple and various. Primas was especially interested in temporal consciousness: the experience of time passing, or the flow of time or the sense that we exist in a fleeting "now" or present. Hence Primas holds that "tenseless physics ... cannot give a complete description of the world" (Primas 2007, p. 30). Here again is a break with the orthodox many-world interpretation of QM, which is *prima facie* completely comfortable with the four-dimensional block view of reality (the block is, however, infinitely foliated like a coral encased in a glass block) and regards the idea of flowing time with deep suspicion.

One puzzle that Primas' views raise is about the complementarity between mind and matter. As noted, complementarity would suggest that mind and matter are co-fundamental. Yet Primas' own pattern-based metaphysics would tend to give a premier role to the mind of the experimenter. This subjective element or choice and perspective seems to be the, or at least a, ground for the emergence of the physical world. Primas wrote that "for a conceptually clean specification of the initial conditions of physical experiments, the homogeneous parameter time of physics has to be complemented by a time with nowness" (Primas 2007, p. 29) and this too might suggest that the experiential side of reality has a metaphysical primacy.

The puzzle is deepened by the fact that at least once when discussing the relation between the physical and mental aspects of the world, Primas denies they are truly fundamental features. For he writes that (Primas 2007, p. 30)

> the tensed domain is supposed to contain the mental domain, while the tenseless domain refers to first principles describing matter and energy. However, the tenseless domain is not identical with physics, it more resembles Plato's non-temporal world of immutable ideas.

This is perplexing because the complementarity of mind and matter would suggest that they are ontologically on a par and co-fundamental. If there is a domain beyond or below that of the material world (Plato's non-temporal world, that is) then the mental too will fail to be fundamental. In philosophy, this worry marks the division between dual-aspect theories and so-called neutral monism.[18] The latter posits a kind of reality which underpins both mind and matter, which can continue to stand in a complementary relation to each other though they lose their status as truly fundamental features of reality. It is pure speculation to attribute either of these views to Primas. It is a great shame that he did not have more time to develop his thoughts on this.

## 6   The Philosophical Legacy of Hans Primas

Primas was of course famous first and foremost as a chemist and quantum chemistry theorist. But it is important to note his philosophical contributions. He made a host of interesting contributions to the philosophy of science. In particular, he was an important force in the revival of the philosophy of chemistry resisting as he did the easy claims that physics had revealed how to reduce chemistry to basic quantum mechanics.

But for me the more interesting aspect of Primas' work goes beyond the philosophy of science. He was not afraid to extend his thought into the metaphysical implications of his views and what he took to be the deep philosophical lessons we should draw from the mysteries of quantum mechanics. Also, although he spent his life as a working scientist, he always resisted an easy or complacent physicalist scientistic vision of the world and opted for an always provisional but audacious embrace of a much richer view of reality. The battle over how narrow a view of reality is acceptable is ancient and still raging. Scientific thinkers such as Primas who marry technical sophistication, deep scientific knowledge and openness to metaphysical speculation are vital warriors helping to keep alive rich and open avenues of thought.

## References

Aiton, E. (1972): *The Vortex Theory of Planetary Motions*, Macdonald, London.
Albert, D. (2015): Probability in the Everett picture. In *After Physics*, Harvard University Press. Cambridge, pp. 161–178.
Atmanspacher, H. and Primas, H. (2003): Epistemic and ontic quantum realities. In *Time, Quantum and Information*, ed. by L. Castell and O. Ischebeck, Springer, Berlin, pp. 301–321.
Badash, L. (1972): The completeness of nineteenth-century science. *Isis* **63**, 48–58.
Banks, E. (2014): *The Realistic Empiricism of Mach, James, and Russell: Neutral Monism Reconceived*, Cambridge University Press, Cambridge.
Barrett, J. (2011): Everett's pure wave mechanics and the notion of worlds. *European Journal for Philosophy of Science* **1**, 277–302.

---

[18]Neutral Monism was famously espoused both by William James and Bertrand Russell as well as Ernst Mach. For an excellent historical overview and modern development of the view see Banks (2014).

Bell, J. (1987): *Speakable and Unspeakable in Quantum Mechanics*, Cambridge University Press, Cambridge.

Berryman, S. (2009): *The Mechanical Hypothesis in Ancient Greek Natural Philosophy*, Cambridge University Press, Cambridge.

Bohm, D. (1952): A suggested interpretation of the quantum theory in terms of "hidden" variables. I. *Physical Review* **85**, 166–179.

Bohm, D., and Hiley, B. (1993): *The Undivided Universe: An Ontological Interpretation of Quantum Mechanics*, Routledge, London.

Born, M. (1926): Quantenmechanik der Stossvorgänge. *Zeitschrift für Physik* **38**, 803–827. Reprinted in translation as "On the Quantum Mechanics of Collisions" in *Quantum Theory and Measurement*, ed. by J. Wheeler and W. Zurek, Princeton University Press, Princeotn 1983, pp. 52–61.

Broad, C.D. (1925): *Mind and Its Place in Nature*, Routledge and Kegan Paul, London.

Chalmers, A. (2001): Maxwell, mechanism, and the nature of electricity. *Physics in Perspective* **3**, 425–438.

Dennett, D. (1991): Real patterns. *Journal of Philosophy* **88**, 27–51. Reprinted in *Dennett's Brainchildren: Essays on Designing Minds*, MIT Press, Cambridge 1998.

d'Espagnat, B. (1999): Concepts of reality: Primas' nonstandard realism. In *On Quanta, Mind, and Matter*, ed. by H. Atmanspacher *et al.*, Kluwer, Dordrecht, 249–272.

Deutsch, D. (1999): Quantum theory of probability and decisions. *Proceedings of the Royal Society of London A* **455**, 3129–3137.

Dirac, P.A.M. (1929): Quantum mechanics of many-electron systems. *Proceedings of the Royal Society of London. Series A* **123**, 714–733.

Dyson, F. (2007): Why is Maxwell's theory so hard to understand? In *Antennas and Propagation*, IEEE/IET Electronic Library, pp. 1–6.

Eliasmith, C. (2000): Is the brain analog or digital? *Cognitive Science Quarterly* **1**, 147–170.

Everett, H. (1957): "Relative state" formulation of quantum mechanics. *Reviews of Modern Physics* **29**, 454–462.

Everett, H. (1973) The theory of the universal wave function. In *The Many-Worlds Interpretation of Quantum Mechanics*, ed. by B. DeWitt and N. Graham, Princeton University Press, Princeton, pp. 3–140.

Finkelstein, D. (1995): Finite physics. In *The Universal Turing Machine. A Half-Century Survey*, ed. by R. Herken, Springer, Wien, pp. 323–347.

Freeth, T., Bitsakis, Y., *et al.* (2006): Decoding the ancient Greek astronomical calculator known as the Antikythera Mechanism. *Nature* **444**, 587–591.

Gangopadhyaya, M. (1981): *Indian Atomism: History and Sources*, Humanities Press, New York.

Gardner, M. (1970): The fantastic combinations of John Conway's new solitaire game "life". *Scientific American* **223**, 120–123.

Greaves, H. (2004): Understanding Deutsch's probability in a deterministic multiverse. *Studies in History and Philosophy of Science B: Studies in History and Philosophy of Modern Physics* **35**, 423–456.

Gregory, J. (1931): *A Short History of Atomism: From Democritus to Bohr*, A. & C. Black, London.

Hameroff, S., and Penrose, R. (1996): Conscious events as orchestrated space-time selections. *Journal of Consciousness Studies* **3**(1), 36–53.

Hensen, B., Bernien, H., *et al.* (2015): Loophole-free Bell inequality violation using electron spins separated by 1.3 kilometres. *Nature* **526**, 682–686.

Holland, P. (1993): *The Quantum Theory of Motion: An Account of the de Broglie-Bohm Causal Interpretation of Quantum Mechanics*, Cambridge University Press, Cambridge.

Joos, E., Zeh, H.D., *et al.* (2003): *Decoherence and the Appearance of a Classical World in Quantum Theory*, Springer, Berlin.

Kent, A. (2010): One world versus many: The inadequacy of Everettian accounts of evolution, probability, and scientific confirmation. In *Many Worlds? Everett, Quantum Theory and Reality*, ed. by S. Saunders, J. Barrett, A. Kent and D. Wallace, Oxford University Press, Oxford, pp. 307–354.

Ladyman, J., Ross, D., *et al.* (2007): *Every Thing Must Go: Metaphysics Naturalized*, Oxford University Press, Oxford.

Laplace, P.-S. (1825/2012): *Pierre-Simon Laplace: Philosophical Essay on Probabilities*, Springer., Berlin. Translated from the 5th French edition of 1825 by A. Dale, with notes by the translator.

Maudlin, T. (2010): Can the world be only wavefunction? In *Many Worlds? Everett, Quantum Theory and Reality*, ed. by S. Saunders, J. Barrett, A. Kent and D. Wallace, Oxford University Press, Oxford, pp. 121–141.

Maxwell, J.C. (1890/1965): *The Scientific Papers of James Clerk Maxwell*, ed. by W.D. Niven, Dover, New York.

McLaughlin, B. (1992): The rise and fall of British emergentism. In *Emergence or Reduction*, ed. by A. Beckermann, H. Flohr and J. Kim, deGruyter, Berlin, pp. 49–93.

Meier, C.A., ed. (2001): *Atom and Archetype. The Pauli/Jung Letters 1932–1958*, Princeton University Press, Princeton.

Mill, J.S. (1843/1963): A system of logic. In *The Collected Works of John Stuart Mill, Vols. 7-8*, University of Toronto Press, Toronto.

Moravec, H. (1988): *Mind Children: The Future of Robot and Human Intelligence*, Harvard University Press, Cambridge.

Mourelatos, A. (1986): Quality, structure, and emergence in later pre-Socratic philosophy. In *Proceedings of the Boston Area Colloquium in Ancient Philosophy* **2**, 127–194.

Murdoch, D. (1989): . *Niels Bohr's Philosophy of Physics*, Cambridge University Press, Cambridge.

Nagel, E. (1961): *The Structure of Science*, Harcourt Brace and Co., New York.

Nagel, T. (1974): What is it like to be a bat? *Philosophical Review* **83**, 435–450. (This article is reprinted in many places, notably in Nagel's *Mortal Questions*, Cambridge University Press, Cambridge 1979.)

Neumann, John von (1932/1955): *Mathematical Foundations of Quantum Mechanics*, Princeton University Press, Princeton.

Newton, I. (1687/1999): *The Principia: Mathematical Principles of Natural Philosophy*, ed. by I.B. Cohen and A. Whitman, University of California Press, Los Angeles.

Newton, I. (1730/1979): *Opticks, or, A Treatise of the Reflections, Refractions, Inflections and Colours of Light*, Dover, New York.

Newton, I. (2004): *Isaac Newton: Philosophical Writings*, ed. by A. Janiak, Cambridge University Press, Cambridge.

Normore, C. (2007): The matter of thought. In *Representation and Objects of Thought in Medieval Philosophy*, ed. by H. Lagerlund, Ashgate, Alderslot, pp. 117–134.

Pauli, W. (1952/1994): The influence of archetypal ideas on the scientific theories of Kepler. In *Wolfgang Pauli. Writings on Physics and Philosophy*, ed. by C. Enz and K. von Meyenn, Springer, Berlin, pp. 219–280.

Poh, H.S., Joshi, S.K., *et al.* (2015): Approaching Tsirelson's bound in a photon pair experiment. *Physical Review Letters* **115**(18), 180408.

Primas, H. (1981/2013): *Chemistry, Quantum Mechanics and Reductionism*, Springer, Berlin.

Primas, H. (1994): Endo-and exo-theories of matter. In *Inside Versus Outside*, ed. by H. Atmanspacher and G. Dalenoort, Springer, Berlin, pp. 163–193.

Primas, H. (1995): Realism and quantum mechanics. In *Logic, Methodology and Philosophy of Science*, ed. by D. Prawitz, D. Westerstahl, and B. Skyrms, Elsevier, Amsterdam, pp. 609–631.

Primas, H. (1998): Emergence in exact natural science. *Acta Polytechnica Scandinavica* **91**, 83–98.

Primas, H. (2003): Time-entanglement between mind and matter. *Mind and Matter* **1**, 81–119.

Primas, H. (2007): Non-Boolean descriptions for mind-matter problems. *Mind and Matter* **5**(1), 7–44.

Pyle, A. (1997): *Atomism and Its Critics: From Democritus to Newton*, Thoemmes Press, Bristol.

Schaffer, S. (2000): Fin de siècle, fin des sciences. *Réseaux* **18**(100), 215–247.

Shimony, A. (1999): Holism. In *On Quanta, Mind, and Matter*, ed. by H. Atmanspacher *et al.*, Kluwer, Dordrecht, pp. 233–248.

Siegel, D.M. (2003): *Innovation in Maxwell's Electromagnetic Theory: Molecular Vortices, Displacement Current and Light*, Cambridge University Press, Cambridge.

Tegmark, M. (2000): Importance of quantum decoherence in brain processes. *Physical Review E* **61**, 4194–4206.

Timpson, C.G. (2008): Quantum Bayesianism: A study. *Studies in History and Philosophy of Science Part B: Studies in History and Philosophy of Modern Physics* **39**, 579–609.

Wallace, D. (2007): Quantum probability from subjective likelihood: Improving on Deutsch's proof of the probability rule. *Studies In History and Philosophy of Science Part B: Studies In History and Philosophy of Modern Physics* **38**, 311–332.

Wallace, D. (2012): *The Emergent Multiverse: Quantum Theory According to the Everett Interpretation*, Oxford University Press, Oxford.

Wallace, D. (2013): A prolegomena to the ontology of the Everett interpretation. In *The Wave Function: Essays on the Metaphysics of Quantum Mechanics*, ed. by A. Ney and D. Albert, Oxford University Press, Oxford, pp. 203–222.

# Contextual Emergence of Deterministic and Stochastic Descriptions

Robert C. Bishop and Peter beim Graben

**Abstract** Hans Primas laid the groundwork for contextual emergence and also had a long-standing interest in issues of stochasticity and determinism and their consequences. In this contribution we describe contextual emergence and then turn to the question of whether determinism and stochasticity could be regarded as contextually emergent notions. In a first step we demonstrate that the conventional concept of determinism is not fully contained in the fundamental description of dynamical systems but requires some contextual stability condition for the emergence of unique trajectories. Second, we discuss mathematical dilation techniques of deterministic systems for the contextual emergence of stochastic descriptions. Finally, the emergence of deterministic "mean field" descriptions from stochastic Markov processes illustrates another contextual aspect of the nature of determinism. We discuss our results regarding contextual determinism and stochasticity in the framework of relative onticity and indicate its potential relevance for the freewill-determinism debates.

## 1 Introduction

The intricate relationships between determinism and stochasticity, between reduction and emergence and between causality and freedom were of deep interest to Hans Primas during much of his scientific career (Primas 1977, 1981, 1998, 2002, 2009). In an early paper, "Theory reduction and non-Boolean theories", he wrote (Primas 1977, p. 283):

> In non-classical theories with a non-Boolean propositional calculus a *restriction* of the domain of discourse can lead to the *emergence of novel properties* and to the appearance of new phenomena. ... The meaningful patterns and the function of a complex system

R.C. Bishop (✉)
Physics Department, Wheaton College, Wheaton, IL, USA
e-mail: robert.bishop@wheaton.edu

P. beim Graben
Bernstein Center for Computational Neuroscience, Humboldt University,
Berlin, Germany
e-mail: prbeimgraben@googlemail.com

© Springer International Publishing Switzerland 2016
H. Atmanspacher and U. Müller-Herold (eds.), *From Chemistry to Consciousness*, DOI 10.1007/978-3-319-43573-2_6

(e.g., a flower) are intrinsically contained in the fundamental description but they manifest themselves only in a theoretical description having a very restricted domain of discourse. By restricting the domain of a fundamental theory, these phenomena can be derived. Such a derivation has to include a historical perspective, e.g. evolutionary processes, because the restriction of the deterministic dynamics of the universal theory to a subtheory induces as a rule a nondeterministic stochastic dynamics.

This short quotation addresses some of the most pertinent problems in the philosophy of science: What is a "fundamental description" of a scientific problem? What is meant by a "restricted domain of discourse" of a scientific theory? In which sense can emergent "novel properties" be derived from such a restriction? And what is the ontological difference between "deterministic dynamics" and induced "stochastic dynamics"? These very questions can be tackled within the current research program of *contextual emergence* that has been pioneered by Primas (1998), later established by Bishop and Atmanspacher (2006) and Atmanspacher and Bishop (2007) and further developed by others (Atmanspacher and Bishop 2007; Bishop 2008, 2010a; beim Graben et al. 2009; Atmanspacher and beim Graben 2009; beim Graben 2011, 2014, 2016; beim Graben and Potthast 2012; Jordan and Ghin 2006).

In this contribution we discuss the relationship between deterministic and stochastic descriptions of physical systems in the light of contextual emergence. We reformulate Primas' claim from the quotation above that stochastic descriptions could be contextually emergent from an underlying deterministic description. This is trivially the case under the assumption of a deterministic ontology yielding a globally deterministic physics from which stochastic descriptions arise under ignorance. Mathematical dilation techniques (Misra et al. 1979; Courbage and Misra 1980; Goldstein et al. 1981; Goodrich et al. 1986; Antoniou and Gustafson 1993; Misra 2002; Gustafson 2002) can lead to emergent stochasticity from an underlying determinism under the condition that the deterministic dynamics is a microscopically chaotic Kolmogorov flow. We show that this requirement can be regarded as a contextual stability condition for the contextual emergence of stochasticity.

However, the converse is also possible: deterministic "mean field" descriptions could also emerge from an underlying stochastic Markovian description by a suitable restriction of the discourse domain. Moreover, we demonstrate that the conventional notion of determinism in terms of unique trajectories is not fully contained in a fundamental description of dynamical systems but is emergent under a suitable contextual restriction.

## 2   Contextual Emergence in Physics

In the quotation above, Primas refers to "non-classical theories with a non-Boolean propositional calculus", and to properties that are "intrinsically contained in the fundamental description". Therefore, his "domain of discourse" is essentially that of quantum theory here. Moreover, the "fundamental description" of "intrinsic" properties refers to the important distinction between ontic and epistemic state descriptions to which Primas also contributed much (Primas 1990a, 1994, see also Atmanspacher's contribution in this volume).

An *ontic state* describes all properties of a state "the way it is" apart from any epistemic access or ignorance. Ontic states belong to individual descriptions with an important special case being the deterministic descriptions of point-particle states and observables in classical mechanics. In contrast, an *epistemic state* describes our knowledge of a physical system's properties gained through our access to such properties based on particular measurement devices and pattern matching routines (Primas 1977, 1981, 1998). Epistemic states belong to statistical descriptions with an important special case being those states and observables describable in terms of probability distributions or density operators.

In its algebraic codification (cf. Primas 1981; Haag 1992), the intrinsic properties of a physical system are described by a C*-algebra of *intrinsic* observables. Such C*-algebras are equipped with a *strong norm topology* which defines convergence and closure properties. Reference to experimentally accessible properties requires a larger W*-algebra of *contextual* observables. A W*-algebra in quantum mechanics is a C*-algebra, usually with a Hilbert space as a pre-dual. This Hilbert space allows the definition of quantum theoretical expectation values as scalar products $\langle \psi | A | \psi \rangle$ (with $A$ being an element of the W*-algebra and $\psi$ belonging to the Hilbert space) and hence defines a weak topology,[1] which is defined through a scalar product in a function Hilbert space.

Now, for a given C*-algebra of intrinsic observables many possible W*-algebras of contextual observables can be constructed that are not (unitarily) equivalent with each other. Such a construction can be carried out along the lines of the Gel'fand-Naimark-Segal (GNS) theorem (cf. Primas 1981; Haag 1992) from a suitably chosen *reference state* taken from the dual of the C*-algebra which is the *ontic state space* of the system. This reference state defines a particular restrictive *context* and, through the GNS-constructed W*-algebra, a *contextual topology*. For a "pure" quantum system the fundamental C*-algebra is factorial and does not contain observables that commute with every other observable (except the identity). The algebra therefore describes a purely non-classical system with a "non-Boolean propositional calculus". Interestingly, the larger W*-algebra may contain a non-empty center, containing commuting and therefore classical observables that are described by a Boolean propositional calculus. Quoting Primas (1981, p. 325, his italics):

> A point of view relative to which a partial description of the world can be given will be called a context. More precisely, *we define a context as a part of the world which is singled out by a well-defined set of prior conceptions whose ontological status is amenable to the application of classical two-valued logic.*

---

[1]The difference between strong and weak topologies can be nicely illustrated by means of series expansions (beim Graben 2016): On the one hand, the Taylor series of a function converges uniformly within its convergence radius; uniform convergence is an example of convergence in a strong topology. On the other hand, the Fourier series of a function converges only in the quadratic norm $L^2$, illustrating a weak topology.

In this sense, novel classical properties can be regarded as *contextually emergent* from a restriction of the domain of discourse of a fundamental theory. Interestingly, globally non-Boolean descriptions that must be restricted to partial Boolean ones are not exclusive to quantum systems. They have been described in the framework of *operational statistics* for generic (classical and quantum) systems by Foulis and co-workers (Foulis and Randall 1972; Randall and Foulis 1973; Foulis 1999; Greechie 1968). In a similar vein, beim Graben and Atmanspacher (2006, 2009) introduced the *epistemic quantization* of classical dynamical systems by means of coarse-gainings and symbolic dynamics.

Contextual emergence was originally introduced to capture the relations between different levels of description in physics (Bishop and Atmanspacher 2006) but it also captures the relations among non-hierarchically structured physical domains (Bishop 2010a). As mentioned above, Primas (1977, 1998) pioneered some of the ideas leading to contextual emergence.

With respect to descriptions, we say that the inter-level relation of *contextual emergence* holds when a given lower-level description contains some necessary but no sufficient conditions for a description at a related higher-level. The missing sufficient conditions for emergent properties are found as higher-level contexts that can be implemented as *stability conditions* at the lower level (Atmanspacher and Bishop 2007). They are required for the identification and persistence of system states at the higher level which are called *macrostates*.

While *microstates*, as referents of an individual, ontic, lower-level description, are pure states in the system's ontic state space, emergent macrostates must persist under various kinds of changes and perturbations at a statistical, epistemic, higher-level description. As a consequence, macrostates are generally non-pure and dispersive (Haag et al. 1974; Atmanspacher and Bishop 2007). The required stability conditions give rise to a contextual topology on the lower-level states, based on a context only available at the higher level. This contextual topology leads to the emergence of new contextual observables at the higher level of description.

One way to see how the necessary and sufficient conditions at the two different levels play a role is through *control hierarchies* (Pattee 1973; Primas 1981). In a control hierarchy, the components or subsystems may provide some necessary conditions for the existence of larger-scale structures; but the larger-scale structures, while never violating the law-like relations among the components and subsystems, constrain their behavior. An example is fluid convection where the molecules provide necessary conditions for the existence of convection cells while the cells restrict or channel the motions of the fluid molecules (Bishop 2012).

An important example of a contextually emergent observable is temperature in thermodynamics. This observable arises through a transition from (point-particle) microstates in either classical or quantum mechanics to an ensemble description in statistical mechanics, which in turn can be extended to thermodynamics by implementing a thermodynamical context.

Let us consider the transitional steps for classical mechanics here. The first one involves going from an individual description in terms of system microstates, represented pointwise, to a statistical ensemble description associated with the total kinetic energy of a system of $N$ particles. A statistical state is given through probability densities defined over regions of the space of microstates. Arbitrary statistical states are in general not dispersion-free, meaning that the computed expectation values for lower-level observables such as velocity yield different results depending on the particular realization of a given statistical state. Thus, statistical states are generally ill-suited for higher-level descriptions in thermodynamics.

The second transition step is often glossed in textbooks and philosophical discussions (e.g., Rorty 1965; Levine 1983) with statements such as that the temperature of a gas equals (or is proportional to) the mean kinetic energy of the molecules constituting the gas. Although this is not wrong, it misses key points of a situation that is actually more subtle: The higher-level observable temperature arises through implementing the contextual condition of thermodynamic equilibrium defined through the zeroth law of thermodynamics in a suitable thermodynamic or continuum limit (Takesaki 1970; Compagner 1989).

The notion of thermodynamic equilibrium does not exist at the level of statistical mechanics. It represents a stability condition at the level of thermodynamics that is implemented through a contextual topology leading to a distinguished set of statistical states, the Kubo-Martin-Schwinger (KMS) states (Kubo 1957; Martin and Schwinger 1959).[2] The KMS condition characterizes the structural stability of such states against local perturbations and implements three crucial stability conditions at the lower level (Haag et al. 1967, 1974; Primas 1998; Bishop and Atmanspacher 2006):

- Expectation values and higher statistical moments of observables computed for KMS states do not change over time (i.e., they are stationary just as thermodynamic observables are in thermal equilibrium).
- KMS states are structurally stable under small perturbations of relevant parameters (i.e., they are ergodic).
- KMS states have no memory of temporal correlations (i.e., they are mixing).

As a result KMS states are approximately dispersion-free macrostates. They represent the lower-level correspondence of higher-level thermal equilibrium states.

The discussed KMS stability criteria can be generalized to coarse-grained descriptions of dynamical systems in the framework of stochastic Markov chains. These systems must be stationary, ergodic and mixing for the contextual emergence of the resulting symbolic dynamics. As a consequence, such Markov chains are provided by generating Kolmororov partitions (Atmanspacher and beim Graben 2007).

---

[2]More rigorously speaking, KMS states arise as local restrictions of a global vacuum state in quantum field theory. This entails that quantum statistical mechanics is needed to derive them—classical statistical mechanics is not enough.

## 3   Determinism in Physics

The usual textbook characterization of determinism is: If the initial and boundary
conditions for a model are selected, then the solutions for the model equations are
fixed for all times. This conception of determinism can be made more precise though.

Determinism is usually understood to be a property of ontic descriptions of states
and observables, whereas stochasticity is usually understood as a property of epistemic descriptions of states and observables (Atmanspacher 2002). In the previous
context of $C^*$-algebraic dynamical systems, a deterministic dynamics is described as
a one-parameter group (or likewise semigroup) of automorphisms $U_t, t \in \mathbb{R}$, such
that $A_t = U_t^* A_0 U_t$ is the observable $A$ after time $t$ given its initial condition $A_0$. In
this sense, ontic time evolution is simply mediated by unitary transformations.

While ontic states are unobservable, it is the ontic level of description where first
principles and universal laws that are deterministic and time-reversible exist. Such
principles and laws cannot be obtained at the epistemic level. Nevertheless, it is
possible to rigorously GNS-construct proper epistemic descriptions from an ontic
description given enough information about the empirically given situation (Primas
1994, 1998). The question of which is the "right" description to use depends on
the proper description of the empirically relevant context (e.g., a definite set of
questions to be addressed, abstracting away irrelevant details). These contexts can
be mathematically specified through contextual topologies. For more details and
examples, see Atmanspacher (2002).

The mathematical models of classical point-particle mechanics have provided the
clearest picture of deterministic descriptions. Three properties of these descriptions
turn out to be particularly important (Bishop 2002, 2005):

- *Differential dynamics*: A state of a system at any given time is related to a state
  at any other time by a nonprobabilistic algorithm. The algorithm could be in the
  form of differential, difference, integral or integro-differential equations (among
  other possibilities, e.g., variational principles such as Hamilton's principle of least
  action). The requirement of non-stochastic equations arises from the restriction
  to ontic descriptions. In particular, a continuous classical dynamical system is
  described by a point $x$ in a $d$-dimensional *space of ontic (pointwise represented)
  states* $\Omega \subset \mathbb{R}^d$ and by a continuous vector field $F : \Omega \times \mathbb{R} \to T\Omega$ mapping an
  ordered pair of a point $x \in \Omega$ and parameter time $t \in \mathbb{R}$ into the tangent bundle
  $T\Omega$ of the manifold $\Omega$, such that the temporal evolution of the system obeys the
  differential equation

$$\dot{x}(t) = F(x, t) .                                                   \tag{1}$$

- *Unique evolution*: A given state is always followed (preceded) by the same history
  of state transitions (restriction to either forwards or backwards time directions
  yields irreversible forms of time evolution). Given the same specification of initial
  and boundary conditions, a mathematical model will undergo the same history
  of transitions from state to state. This property is not redundant with differential dynamics. For instance by either underdetermining or overdetermining the

conditions for differential equations, uniqueness and/or existence may be lost. Differential dynamics allows a great deal of freedom in choosing algorithms, including algorithms lacking uniqueness properties (see below). Mathematically, unique evolution essentially states that for a given initial condition $x_0 \in \Omega$ there exists one and only one solution of Eq. (1) such that

$$x(t) = \Phi^t(x_0) \tag{2}$$

with a one-parameter automorphism group (or semigroup) representation on the phase space, called the flow $\Phi^t$.

- *Value determinateness*: Any state can be described with arbitrarily small (nonzero) error. While this is a common assumption in textbook discussions of point-particle mechanics, it should be pointed out that Schrödinger evolution is an example where differential dynamics and unique evolution hold while value determinateness fails. This suggests that it is possible to have interval-valued states and maintain a form of determinism, but this involves a change in the way states are described (Bishop 2005). Moreover, value determinateness can be related to *strong causality* in the sense that small differences among initial conditions remain small during the dynamics. Mathematically, this relates to Lipschitz continuity of the flow:

$$||\Phi^t(x) - \Phi^t(y)|| \leq K||x - y|| \tag{3}$$

for $x, y \in \Omega$, where $K$ is the Lipschitz constant. Then, every trajectory starting in a small set of initial conditions $B_0 \subset \Omega$ is contained in a tube whose time-slices are roughly congruent to the initial $B_0$.

## 4  Contextual Emergence of Determinism and Stochasticity

We are now ready to discuss the status of determinism from the vantage point of contextual emergence. At first it may seem surprising to think of determinism as contextual rather than universal, but determinism has always had a delicate status in physics (Earman 1986). For example, unique evolution fails to hold for force functions proportional to $\sqrt{v}$ in Newton's equations of motion for all specifications of the initial position and velocity $v$. There are other force functions that lead to violations of unique evolution and these can only be ruled out based on contextual reasons.

Newton's inverse-square gravitational force is one of these forces. The French mathematician Paul Painlevé conjectured that a system of point-particles interacting under Newtonian gravity could accelerate some (perhaps all) particles to spatial infinity within a finite time interval, violating unique evolution. The infinite potential well associated with the inverse-square force is the source of the energy for this acceleration. Painlevé's conjecture was proven in 1992 for a system of as few as five point masses (Zhihong 1992).

## *4.1   Emergent Determinism I*

How might we diagnose the apparent lack of universal determinism? The description of a dynamical system as given above does not indicate anything about the existence or uniqueness of solutions of Eq. (1) for a given initial condition $x_0 \in \Omega$, or even about the existence of a flow $\Phi^t$ or of trajectories. In this sense, differential dynamics as expressed by Eq. (1) is only a necessary condition for unique evolution.

Moreover, it is important to notice that differential dynamics expressed by a differential equation (1) already refers to a given context in terms of a coordinate system for the space $\Omega$. The transition from one coordinate system to another one is generally described through principles of relativity. For example, Newton's second law of motion $m\ddot{x} = F(x)$ is invariant under Galilei transformations, but not under Lorentz transformations. Therefore, it only holds for inertial systems but not for frames of reference moving with a velocity close to that of light or for accelerated frames of reference which are properly described by general relativity. For the latter, the impressed force $F$ must be superimposed with inertia pseudo-forces; a good example for this is the Coriolis force in the earth's atmospheric system. Thus, the general vector field $F$ in Eq. (1) depends on a particular frame of reference and thereby on a context.

In the theory of ordinary differential equations, the existence and uniqueness of solutions is proven under the sufficient condition that the vector field $F$ is locally Lipschitz continuous, i.e., for all times $t \in \mathbb{R}$ and for all points $x, y \in \Omega$:

$$||F(x, t) - F(y, t)|| \leq L||x - y||, \tag{4}$$

with locally different Lipschitz constants $L$. Since Lipschitz continuity provides an upper bound for the distortion of a distance between two points $x, y \in \Omega$ as mediated by the function $F$, we consider this sufficient condition as a *stability criterion* for the contextual emergence of determinism. If Eq. (1) for a given initial condition $x_0 \in \Omega$ satisfies Lipschitz continuity, then it is a well-set initial value problem in the sense that its solutions have the property of unique evolution. Nevertheless, as discussed by Arnold (1988), such sufficient conditions may only hold for a finite (and perhaps very brief) interval of time (e.g., $x(t)$ may become infinite for $t < \infty$, as in case of particles interacting under Newtonian gravity). This means that Eq. (4) may guarantee the existence of unique solutions of Eq. (1) for all times or only for some interval of time.

Since unique evolution is contextually emergent through Lipschitz continuity as a stability criterion, a given initial condition $x_0 \in \Omega$ generates a unique trajectory $T = \{x(t) \in \Omega | \dot{x}(t) = F(x, t) \text{ and } x(0) = x_0\}$. Similarly, we regard the bundle of all trajectories starting at different initial conditions as the phase flow $\Phi^t : \Omega \to \Omega$, such that $x(t) = \Phi^t(x_0)$ with $\Phi$ being a group homomorphism $\Phi^{s+t} = \Phi^s \circ \Phi^t$ ($s, t \in \mathbb{R}$) for invertible dynamical systems.

## 4.2 Emergent Stochasticity

The contextual emergence of stochastic from deterministic descriptions was a focal area in Primas' work. In 1990 he proved how a quantum spin system coupled to the infinite number of degrees of freedom of a quantum harmonic oscillator constituting the surrounding electromagnetic field gives rise to a nonlinear stochastic evolution equation for the magnetic momentum (Primas 1990b). The particular context selected for this derivation is given as a Cauchy-type environment.

Let us consider an epistemic description of either uncertainty or ignorance about initial conditions. This means, initial conditions cannot be prepared exactly but rather as a set $B_0 \subset \Omega$ of non-zero measure with respect to some probability measure $\mu$, thereby relaxing value determinateness. We consider a Kolmogorov probability space $(\Omega, B, \mu)$ over the space $\Omega$ with measure $\mu$ and a Borel $\sigma$-algebra of measurable subsets $B \in B$. In general, uncertainty about initial conditions can be expressed by a probability density function $\rho_0(x)$ over $\Omega$. As the deterministic dynamics obeys Eqs. (1) and (2), the resulting dynamics of probability densities is described by a Frobenius-Perron integral equation (Ott 1993)

$$\rho(x, t) = \int_X \delta(x - \Phi^t(x')) \, \rho_0(x') \, \mathrm{d}x' \tag{5}$$

which can be expressed through a Frobenius-Perron operator $U_t$ (Goldstein et al. 1981; Antoniou and Gustafson 1993):

$$\rho(x, t) = [U_t \rho_0](x) = \rho_0(\Phi^{-t}(x)) . \tag{6}$$

The Frobenius-Perron operator $U_t$, acting upon probability densities, is the adjoint of the Koopman operator $V_t$ (Koopman 1931; Koopman and von Neumann 1932), acting on the algebra of observables of complex-valued functions defined over $\Omega$: $[V_t f](x) = f(\Phi^t(x))$. This can be easily seen by calculating the expectation functional:

$$\begin{aligned}
\langle f \rangle_{\rho_t} &= \int_\Omega f(x)\rho(x, t)\mathrm{d}x = \int_\Omega f(x)[U_t\rho_0](x)\mathrm{d}x \\
&= \int_\Omega [U_t^* f](x)\rho_0(x)\mathrm{d}x = \int_\Omega [V_t f](x)\rho_0(x)\mathrm{d}x.
\end{aligned}$$

A proper stochastic dynamics can be similarly described by a transition function $P_t(A|x)$ which describes the dispersion of a singular probability density focused in a point $x \in \Omega$ to a measurable set $A \in B$ after transition time $t$ (Misra et al. 1979; Goldstein et al. 1981; Antoniou and Gustafson 1993). From this an operator

$$[W_t f](x) = \int_\Omega f(x') P_t(\mathrm{d}x'|x) \tag{7}$$

acting on observables can be defined which describes a stationary Markov process through a semigroup homomorphism. Its corresponding adjoint acting on probability densities, $W_t^*$, leads then to a stochastic dynamics over the same space.

As has been pointed out by Misra et al. (1979), Courbage and Misra (1980), Goldstein et al. (1981), Goodrich et al. (1986), Antoniou and Gustafson (1993), Misra (2002), Gustafson (2002), dynamical systems can be intrinsically random if they are microscopically so unstable or chaotic that the concept of a trajectory is only an epistemically unrealizable ontic ideal. In this case, strong causality (3) of the flow $\Phi^t$ breaks down and the principle of value determinateness becomes even more violated. Hence, any small measurable subset of initial conditions $B_0 \in \mathcal{B}$ spreads out due to the exponential divergence of chaotic trajectories.

The idea of intrinsic randomness can be captured through the notion of a Kolmogorov flow, briefly K-flow (Walters 1981; Goldstein et al. 1981) as follows. Consider a finite partition $\mathcal{P}$ of $\Omega$ into pairwise disjoint measurable sets $P_i, P_j \in \mathcal{P}$, $P_i \cap P_j = \emptyset$, covering the entire space $\bigcup_i P_i = \Omega$. For two partitions, $\mathcal{P}, \mathcal{Q}$, the *partition product* $\mathcal{P} \vee \mathcal{Q}$ is defined as the set of all intersections of any two subsets of $\mathcal{P}, \mathcal{Q}$, respectively, i.e.:

$$\mathcal{P} \vee \mathcal{Q} = \{ P_i \cap Q_j | P_i \in \mathcal{P}, Q_j \in \mathcal{Q} \} . \tag{8}$$

Finally, the concept of *dynamic refinement* is introduced through the pre-images of a partition $\Phi^{-t}(\mathcal{P}) = \{\Phi^{-t}(P_i)|P_i \in \mathcal{P}\}$ as the product partition

$$\mathcal{P}_t = \bigvee_{-\infty < \tau \leq t} \Phi^{-\tau}(\mathcal{P}) . \tag{9}$$

In order for a dynamical system to be a K-flow, the system must possess a so-called K-partition $\mathcal{P}$ such that

- The partition $\mathcal{P}_t$ is finer than $\mathcal{P}_s$ for $t > s$.
- In the limit of infinite future, the finest dynamic refinement of $\mathcal{P}_t$ is the identity partition $\mathcal{I}$ corresponding to the original Borel algebra $\mathcal{B}$ (modulo zero-sets): $\lim_{t \to \infty} \mathcal{P}_t = \mathcal{I}$ (beim Graben and Atmanspacher 2009).
- In the limit of infinite past, we have $\lim_{t \to -\infty} \mathcal{P}_t = \mathcal{T}$, i.e. the trivial partition $\mathcal{T}$ corresponding to the Borel algebra that contains the entire space $\Omega$ (modulo zero-sets).

For time-discrete dynamical systems, these conditions define a generating partition which Atmanspacher and beim Graben (2007) utilized as a stability condition for contextual emergence. This is in close analogy to the dilation approach discussed here (cf. Misra 2002; Gustafson 2002), where a deterministic dynamics, given by the Frobenius-Perron operator $U_t$, and a stochastic dynamics, described by a Markov operator $W_t^*$, can be transformed by a similarity transformation $\Lambda$ such that

$$\Lambda^{-1} W_t^* \Lambda = U_t . \tag{10}$$

This construction is possible under the necessary condition that the Markov process described by $W_t^*$ is mixing such that it approaches an equilibrium distribution under entropy maximization (Misra et al. 1979; Goldstein et al. 1981). On the other hand, the sufficient condition is that the deterministic dynamics $U_t$ is a K-flow (Misra et al. 1979; Goldstein et al. 1981), thereby implementing microscopic chaos as a stability condition.

The similarity transformation in Eq. (10) expresses the fact that stochastic or deterministic descriptions depend on a chosen context of a particular K-partition. Note that this partition simultaneously provides a contextual topology, as topologies can be defined through families of open (or likewise closed) sets of $\Omega$. Finally, the distinction between necessary and sufficient conditions indicates that stochastic descriptions are contextually emergent from deterministic descriptions in the dilation approach.

## 4.3 Emergent Determinism II

Perhaps surprisingly, the converse can be the case as well. Consider a stochastic dynamical system whose Markov operator $W_t^*$ as in Eq. (7) gives rise to a master equation (Kampen 1992)

$$\frac{\partial \rho(x,t)}{\partial t} = \int_\Omega w(x,y)\rho(y,t) - w(y,x)\rho(x,t)\,\mathrm{d}y , \qquad (11)$$

where $w(x,y)$ is the transition rate from state $y$ into state $x$. Computing the expectation of an observable $f$ under the statistical states $\rho(x,t)$,

$$F(t) = \langle f \rangle_{\rho_t} = \int_\Omega f(x)\rho(x,t)\mathrm{d}x , \qquad (12)$$

yields the macroscopic observable, or "mean field", $F(t)$. Its temporal derivative is obtained from the master equation (11),

$$\begin{aligned}
\frac{\mathrm{d}F(t)}{\mathrm{d}t} &= \int_\Omega f(x)\frac{\mathrm{d}\rho(x,t)}{\mathrm{d}t}\,\mathrm{d}x \\
&= \int_\Omega \int_\Omega f(x)[w(x,y)\rho(y,t) - w(y,x)\rho(x,t)]\,\mathrm{d}x\,\mathrm{d}y \\
&= \int_\Omega \int_\Omega [f(y) - f(x)]w(y,x)\rho(x,t)\,\mathrm{d}x\,\mathrm{d}y \\
&= \int_\Omega a_1(f(x))\rho(x,t)\,\mathrm{d}x \\
&= \langle a_1(f(x)) \rangle_{\rho_t}
\end{aligned} \qquad (13)$$

with the jump moment

$$a_m(x) = \int_\Omega (y - x)^m w(y, x) \, dy \, . \tag{14}$$

From (13) we obtain an emergent deterministic macrostate dynamics

$$\frac{dF(t)}{dt} = a_1(F(t)) \tag{15}$$

in two distinguished cases: (1) when the first jump moment $a_1$ is a linear function and, more importantly, (2) when the variances and all higher statistical moments of $F$ are almost vanishing for the solution states of the master equation. In the latter case, $a_1$ can be expanded into a power series with only the linear term contributing to (15). Remarkably, this case is analogous to our paradigmatic example of the contextual emergence of thermal equilibrium states, where thermal KMS macrostates are relatively pure and, hence, almost dispersion-free. As this results from the mixing property of Markov processes we find a sufficient stability condition also for the contextual emergence of deterministic macrostate dynamics.

# 5 Discussion

In the previous sections we have demonstrated that both determinism and stochasticity can be conceived as contextually emergent descriptions of physical systems. Moreover, deterministic and stochastic descriptions for the very same system may apply at different descriptive levels.

An important example for such alternations can be found in the neurosciences (beim Graben 2016): Neuronal dynamics crucially depends on the function of ion channels that are embedded in the membranes of nerve cells. These are protein macromolecules constituting a pore through the cell membrane that is permeable for specific ions. Depending on its environment the pore could open or close, a dynamics that is described by a Markov chain at a high functional level (Hille 2001). On the lowest level, however, the channel must be seen as a quantum object in an entangled state. At this level of description, the channel has no classical properties: there is no shape, no pore, no open or closed state. There is only a quantum state governed by a deterministic unitary Schrödinger dynamics.

According to Primas (1981, 1998), this low-level description breaks down through the singular perturbation expansion of the Born and Oppenheimer (1927) procedure where a nuclear frame and thus the shape of the channel are contextually emergent as classical properties (see Bishop and Atmanspacher 2006 for discussion). This emergence is, though, at the expense of a stochastic dynamics obtained for the nuclear bodies. Approximating this stochastic dynamics in a coarse-grained description of functional open/closed states by a mixing ergodic Markov chain leads to a deterministic evolution equation that is part of the famous Hodgkin-Huxley equations for neural function (Hodgkin and Huxley 1952).

Since alternating deterministic and stochastic descriptions depend on the choice of particular contexts, our findings are at variance with a primordial deterministic ontology yielding a globally deterministic physics from which stochastic descriptions only arise under epistemic ignorance—as is often assumed in the philosophical literature.[3] By contrast, the issue of determinism or stochasticity depends on the choice of a contextual topology for a given descriptive level. Thus, the situation might be better understood in terms of a framework called *relative onticity* (Atmanspacher and Kronz 1999) that is inspired by Quine's *ontologial relativity* (Quine 1969).

Relative onticity emphasizes the ineliminable role of contexts in scientific descriptions. As described above, descriptions can be either ontic or epistemic. Similarly, the same descriptive framework can be construed as either ontic or epistemic, depending on which other framework it is related to. For instance, the movement of individual molecules in a cup of cappuccino will be highly relevant as an ontic feature when analyzing motion, but entirely irrelevant when enjoying the taste of it. Another example: individual $H_2O$ molecules, considered as ontic constituents of water, are not wet. The property of the liquidity of water emerges in a thermodynamic description similar to the way the property of temperature is contextually emergent.

We see something similar for determinism and stochasticity. A description of individual particles where Eq. (4) holds can be regarded as an ontic deterministic description from which a stochastic description of ensembles of particles is contextually emergent as an epistemic description. However, we can also treat this stochastic description as an ontic description from which a higher-level deterministic description is contextually emergent in a mean-field approach. Successive ontic/epistemic levels of description interleave in a hierarchy where, instead of asking which description is more fundamental, we ask which one represents the relevant context. For each context we can then ask under what conditions determinism or stochasticity is an emergent feature.

Making use of relative onticity and inter-level relationships between stochastic and deterministic descriptions may prove helpful for characterizing dynamics in actual-world systems. For instance, in a recent study by Frentz et al. (2015) the population dynamics of three different microbial species were found to be "strongly deterministic." By this term the authors mean that the dynamics of the three species were highly replicable even though there are numerous stochastic factors involved. Although equating replicability with determinism is highly problematic (Earman 1986; Atmanspacher 2002; Primas 2002; Bishop 2005), it is possible, as in the cases of KMS macrostates and the Hodgkin-Huxley equations, to precisely specify the relationship between stochastic and deterministic dynamics at different levels of description and identify stability conditions for deterministic dynamics. Contextual emergence provides the formal framework to do so.

---

[3]See a number of contributions in the *Oxford Handbook of Free Will* Kane (2011) for pertinent examples.

An important implication of the contextuality of determinism is that a universal deterministic physics is untenable. This is in agreement with Primas' argument that experimental physics presupposes the freedom of experimenters to set up a measurement procedure and to deliberately prepare the initial conditions of their experiments (Primas 2002, 2009). For a deterministic system, once these choices are made, the experiment proceeds. If physical determinism were universal, the experimenter's freedom to choose initial conditions would not exist and all of experimental physics would be pointless. While an ontic deterministic description may be appropriate for a particular given situation, an epistemic nondeterministic description is relevant at the level of the experimental scientist's action (for more discussion see Bishop 2010b). A further implication of determinism's contextuality is that questions of determinism versus free will become contextual as well (Bishop and Atmanspacher 2011).

Since the 17th century physics has been our best guide for thinking how universal determinism governs the behavior of systems in nature. However, if determinism is a contextually emergent property, this suggests revisions of our metaphysical assumptions and conceptions. We like to think that Hans Primas would have enjoyed (and critically commented on) such rethinking of determinism and stochasticity.

# References

Antoniou, I.E., and Gustafson, K.E. (1993): From probabilistic descriptions to deterministic dynamics. *Physica A* **197**, 153–166.

Arnold, V.I. (1988): *Geometrical Methods in the Theory of Ordinary Differential Equations*, Springer, New York.

Atmanspacher, H. (2002): Determinism is ontic, determinability is epistemic. In *Between Chance and Choice*, ed. by H. Atmanspacher and R.C. Bishop, Imprint Academic, Thorverton, pp. 49–74.

Atmanspacher, H., and Bishop, R.C. (2007): Stability conditions in contextual emergence. *Chaos and Complexity Letters* **2**(2/3), 139–150.

Atmanspacher, H., and beim Graben, P. (2007): Contextual emergence of mental states from neurodynamics. *Chaos and Complexity Letters* **2**(2/3), 151–168.

Atmanspacher, H., and beim Graben, P. (2009): Contextual emergence. *Scholarpedia* **4**(3), 7997. Accessible at http://www.scholarpedia.org/article/Contextual_emergence.

Atmanspacher, H., and Kronz, F. (1999): Relative onticity. In *On Quanta, Mind, and Matter*, ed. by H. Atmanspacher, *et al.*, Kluwer, Dordrecht, pp. 273–294.

Bishop, R. (2002): Deterministic and indeterministic descriptions. In *Between Chance and Choice*, ed. by H. Atmanspacher and R.C. Bishop, Imprint Academic, Thorverton, pp. 5–31.

Bishop, R.C. (2005): Anvil or onion? Determinism as a layered concept. *Erkenntnis* **63**, 55–71.

Bishop, R.C. (2008): Downward causation in fluid convection. *Synthese* **160**, 229–248.

Bishop, R.C. (2010a): Whence chemistry? *Studies in History and Philosophy of Science Part B* **41**, 171–177.

Bishop, R.C. (2010b): Free will and the causal closure of physics. In *Visions of Discovery: New Light on Physics, Cosmology, and Consciousness*, ed. by R.Y. Chiao *et al.*, Cambridge Univesity Press, Cambridge, pp. 601–611.

Bishop, R.C. (2012): Fluid convection, constraint and causation. *Interface Focus* **2**, 4–12.

Bishop, R.C., and Atmanspacher, H. (2006): Contextual emergence in the description of properties. *Foundations of Physics* **36**, 1753–1777.

Bishop, R.C., and Atmanspacher, H. (2011): Causal closure of physics and free will. In *Oxford Handbook of Free Will*, ed. by R. Kane, Oxford University Press, Oxford, pp. 101–114.

Born, M., and Oppenheimer, R. (1927): Zur Quantentheorie der Molekeln. *Annalen der Physik* **389**, 457–484.

Compagner, A. (1989): Thermodynamics as the continuum limit of statistical mechanics. *American Journal of Physics* **57**, 106–117.

Courbage, M., and Misra, B. (1980): On the equivalence between Bernoulli dynamical systems and stochastic Markov processes. *Physica A* **104**, 359–377.

Earman, J. (1986): *A Primer on Determinism*, Reidel, Dordrecht.

Foulis, D.J. (1999): A half-century of quantum logic. What have we learned? In *Quantum Structures and the Nature of Reality*, ed. by D. Aerts, Kluwer, Dordrecht, pp. 1–36.

Foulis, D.J., and Randall, C.H. (1972): Operational statistics I. Basic concepts. *Journal of Mathematical Physics* **13**, 1667–1675.

Frentz, Z., Kuehn S., and Leibler, S. (2015): Strongly deterministic population dynamics in closed microbial communities. *Physical Review X* **5**(041014), 1–18.

Goldstein, S., Misra, B., and Courtage, M. (1981): On intrinsic randomness of dynamical systems. *Journal of Statistical Physics* **25**, 111–126.

Goodrich, R.K., Gustafson, K., and Misra, B. (1986): On K-flows and irreversibility. *Journal of Statistical Physics* **43**, 317–320.

beim Graben, P. (2011): Naphtas Visionen. Perspektivität in der Naturwissenschaft. In *Post-Physikalismus*, ed. by M. Knaup *et al.*, Alber, Freiburg, pp. 122–141.

beim Graben, P. (2014): Contextual emergence of intentionality. *Journal of Consciousness Studies* **21**(5-6), 75–96.

beim Graben, P. (2016): Contextual emergence in neuroscience. In *Closed Loop Neuroscience*, ed. by A.E. Hady, Elsevier Amsterdam, in press.

beim Graben, P., and Atmanspacher, H. (2006): Complementarity in classical dynamical systems. *Foundations of Physics* **36**, 291–306.

beim Graben, P., and Atmanspacher, H. (2009): Extending the philosophical significance of the idea of complementarity. In *Recasting Reality*, ed. by H. Atmanspacher and H. Primas, Springer, Berlin, pp. 99–113.

beim Graben, P., Barrett, A., and Atmanspacher, H. (2009): Stability criteria for the contextual emergence of macrostates in neural networks. *Network: Computation in Neural Systems* **20**, 178–196.

beim Graben, P., and Potthast, R. (2012): Implementing Turing machines in dynamic field architectures. In *Proceedings of AISB12 World Congress 2012 – Alan Turing*, ed. by M. Bishop and Y.J. Erden, Society for the Study of Artificial Intelligence, Birmingham, pp. 36–40.

Greechie, R.J. (1968): On the structure of orthomodular lattices satisfying the chain condition. *Journal of Combinatorial Theory* **4**, 210–218.

Gustafson, K. (2002): Time-space dilations and stochastic-deterministic dynamics. In *Between Chance and Choice*, ed. by H. Atmanspacher and R.C. Bishop, Imprint Academic, Thorverton, pp. 115–148.

Haag, R. (1992): *Local Quantum Physics: Fields, Particles, Algebras*, Springer, Berlin.

Haag, R., Hugenholtz, N.M., and Winnink, M. (1967): On the equilibrium states in quantum statistical mechanics. *Communications in Mathematical Physics* **5**, 215–236.

Haag, R., Kastler, D., and Trych-Pohlmeyer, E.B. (1974): Stability and equilibrium states. *Communications in Mathematical Physics* **38**, 173–193.

Hille, B. (2001): *Ion Channels of Excitable Membranes*, Sinauer, Sunderland.

Hodgkin, A.L., and Huxley, A.F. (1952): A quantitative description of membrane current and its application to conduction and excitation in nerve. *Journal of Physiology* **117**, 500–544.

Jordan, J.S., and Ghin, M. (2006): (Proto-)Consciousness as a contextually emergent property of self-sustaining systems. *Mind and Matter* **4**(1), 45–68.

van Kampen, N.G. (1992): *Stochastic Processes in Physics and Chemistry*, Elsevier, Amsterdam.

Kane, R., ed. (2011): *Oxford Handbook of Free Will*, Oxford University Press, Oxford.

Koopman, B.O. (1931): Hamiltonian systems and transformations in Hilbert space. *Proceedings of the National Academy of Sciences of the USA* **17,** 315–318.

Koopman, B.O., and von Neumann, J. (1932): Dynamical systems of continuous spectra. *Proceedings of the National Academy of Sciences of the USA* **18,** 255–263.

Kubo, R. (1957): Statistical-mechanical theory of irreversible processes. I. General theory and simple applications to magnetic and conduction problems. *Journal of the Physical Society of Japan* **12,** 570–586.

Levine, J. (1983): Materialism and qualia: The explanatory gap. *Pacific Philosophical Quarterly* **64,** 354–361.

Martin, P., and Schwinger, J. (1959): Theory of many-particle systems. I. *Physical Review* **115,** 1342–1373.

Misra, B. (2002): Transitions from deterministic evolution to irreversible probabilistic processes and the quantum measurement problem. In *Between Chance and Choice*, ed. by H. Atmanspacher and R.C. Bishop, Imprint Academic, Thorverton, pp. 149–163.

Misra, B., Prigogine, I., and Courbage, M. (1979): From deterministic dynamics to probabilistic descriptions. *Proceedings of the National Academy of Sciences of the USA* **76,** 3607–3611.

Ott, E. (1993): *Chaos in Dynamical Systems*, Cambridge University Press, New York.

Pattee, H. (1973): The physical basis and origin of hierarchical control. In *Hierarchy Theory: The Challenge of Complex Systems*, George Braziller, New York, pp. 69–108.

Primas, H. (1977): Theory reduction and non-Boolean theories. *Journal of Mathematical Biology* **4,** 281–301.

Primas, H. (1981): *Chemistry, Quantum Mechanics and Reductionism*, Springer, Berlin.

Primas, H. (1990b): Induced nonlinear time evolution of open quantum objects. In *Sixty-Two Years of Uncertainty*, ed. by A.I. Miller, Plenum, New York, pp. 260–280.

Primas, H. (1990a): Mathematical and philosophical questions in the theory of open and macroscopic quantum systems. In *Sixty-Two Years of Uncertainty*, ed. by A.I. Miller, Plenum, New York, pp. 233–257.

Primas, H. (1994): Endo- and exo-theories of matter. In *Inside Versus Outside*, ed. by H. Atmanspacher, and G.J. Dalenoort, Springer, Berlin, pp. 163–193.

Primas, H. (1998): Emergence in exact natural sciences. *Acta Polytechnica Scandinavica* **91,** 83–98.

Primas, H. (2002): Hidden determinism, probability, and time's arrow. In *Between Chance and Choice*, ed. by H. Atmanspacher and R.C. Bishop, Imprint Academic, Thorverton, pp. 89–113.

Primas, H. (2009): Complementarity of mind and matter. In *Recasting Reality*, ed. by H. Atmanspacher and H. Primas, Springer, Berlin, pp. 171–210.

Quine, W.V. (1969): Ontological relativity. In *Ontological Relativity and Other Essays*, Columbia University Press, New York.

Randall, C.H., and Foulis, D.J. (1973): Operational statistics. II. Manuals of operations and their logics. *Journal of Mathematical Physics* **14,** 1472–1480.

Rorty, R. (1965): Mind-body identity, privacy, and categories. *Review of Metaphysics* **19,** 24–54.

Takesaki, M. (1970): Disjointness of the KMS-states of different temperatures. *Communications in Mathematical Physics* **17,** 33–41.

Walters, P. (1981): *Introduction to Ergodic Theory*, Springer, Berlin.

Zhihong, X. (1992): The existence of noncollision singularities in Newtonian systems. *Annals of Mathematics* **135,** 411–468.

# Aspects of Algebraic Quantum Theory: A Tribute to Hans Primas

Basil J. Hiley

**Abstract** This paper outlines the common ground between the algebraic approach to quantum phenomena proposed by Hans Primas and the ideas lying behind David Bohm's notion of the implicate and explicate order. The latter emerged from what he called "an algebraic description of structure- process" which, in terms of formal logic, was a way to study the relation between a non-Boolean (implicate) quantum logic and its Boolean (explicate) projections. We show that in the implicate order, we have two time-evolution equations, one involving a commutator, which is essentially Heisenberg's equation of motion, and the other involving an anti-commutator or Jordan product. Explicate orders emerge from projections into, or shadows on, Boolean sub-structures, a process that Primas has likened to "pattern recognition". These projections produce equations that form the basis of what has been called the de Broglie–Bohm interpretation of quantum mechanics. By exploiting the properties of the orthogonal Clifford algebras, this model has been generalized to include relativistic systems with spin, giving a novel insight into the whole approach.

## 1 Introduction

### 1.1 The Common Ground

It is a privilege to be invited to contribute to this volume dedicated to Hans Primas whose work on the foundations of quantum theory has had a strong influence on my own thinking on the subject. I first came across his ideas on algebraic quantum mechanics in a bound manuscript entitled *Quantum Mechanical System Theory* in David Bohm's room at Birkbeck College in 1977, one year later published by Primas

B.J. Hiley (✉)
Physics Department, University College, London, UK
e-mail: b.hiley@bbk.ac.uk

B.J. Hiley
Birkbeck College, London, UK

© Springer International Publishing Switzerland 2016
H. Atmanspacher and U. Müller-Herold (eds.), *From Chemistry to Consciousness*, DOI 10.1007/978-3-319-43573-2_7

and Müller-Herold (1978). The manuscript was to prove invaluable for my thinking about quantum theory.

I had been working with David Bohm trying to develop a new way of thinking about quantum theory based on a process philosophy, which we formulated in terms of an algebraic structure along the lines of the original proposals of Born and Jordan (1925).The idea of using an algebraic structure to describe process has an even longer history going back to Hamilton (1967), Grassmann (1894, 1995), and Clifford (1882). But for one reason or another it fell into disrepute, in spite of its use by Eddington (1936).

Fortunately today the notion of process as fundamental is undergoing a revival, particularly with the appearance of category theory especially in the hands of Abramsky and Coecke (2004) and Coecke (2005), who use the theory in the context of quantum mechanics, explaining in greater detail their motivations for using a process approach. In this paper I prefer to motivate the algebraic theory along lines that are more closely linked with the approach developed by Primas. Indeed it was his manuscript that first drew my attention to the advantages of the more general C*-algebraic approach, an algebraic structure that I was completely unaware of at the time.

My interests in an algebraic approach had already been aroused by Penrose (1971) twistor theory, a generalization of the Dirac-Clifford algebra introduced by Dirac to describe the relativistic electron. At the time Penrose was in the mathematics department at Birkbeck and, together with Bohm, we would meet regularly for seminars that were concerned with the possibility of developing quantum space-time structures, a radical idea that we thought necessary in order to unite quantum theory with general relativity.

Penrose (1971) was also exploring the possibility of developing a description based on a discrete spin network, thus avoiding the need to assume an a priori given space-time continuum (Penrose 1967). This idea of a network structure fitted in very nicely with the topic of my Ph.D. thesis, although that was in a very different field.

My thesis involved investigating certain aspects of the Ising model used in the study of cooperative phenomena in solid state physics. The simple model that I was exploring involved determining the thermodynamics of a many-particle lattice system with nearest neighbour interactions. It was based on a method of finite clusters, using an idea first proposed by Domb and Hiley (1962). The evaluation of the partition function, and hence the thermodynamical properties, necessitated developing a technique for embedding finite graphs in regular tessellations. What I noticed was that some of these properties, essentially combinatorial in nature, depended only on the dimensionality of the embedding space and not on the detailed structure of the tessellation. In other words, simply by *counting* embeddings, one could determine the dimensionality of the embedding space (Hiley et al. 1977). Later I became aware of the fact that the partition function could be obtained much more simply using an algebraic approach used in knot theory. This approach was described by Kauffman (2001, p. 373) who illustrated the technique on small clusters.

The phrase "quantum space-time" was a generic term to refer to any structure that did not take a continuum of points as fundamental, but rather the points were assumed to emerge from a deeper structure. That was, in fact, the idea behind the Penrose twistor which is used to describe a complex of light rays whose intersections define the points of space-time. He also found that congruences of light rays twisting around each other could be used to define sets of "extended points" which he hoped would avoid some of the singularities that plague quantum electrodynamics.

But surely finding partition functions of a spin lattice is a long way from the problems of developing a quantum space-time? Not so—because it turns out that the algebraic techniques lying behind both twistors and the algebraic evaluation of partition functions are closely related to the seminal work of Vaughn Jones (1986) on von Neumann algebras. In a remarkable paper, he showed the connection between these algebras and the combinatorial properties of knots which, as we have already remarked, lie at the heart of the techniques involved in evaluating the partition function of finite clusters of spin systems. The connection becomes even more suggestive when it is realized that the Onsager exact solution (Onsager 1944) for the two-dimensional Ising model involves a Clifford algebra, an algebra that is one example of a von Neumann algebra. Note also that these algebras are the very algebras that Penrose (1971) used to construct his twistors. However all these ideas were then yet to unfold in the future.


## 1.2   Structure-Process

In those early days, Bohm (1965, 1971) was developing his notion of "structure-process" which emphasized the relationships, order and structure of a network of elementary processes. Not relations that could be embedded in the Cartesian order of points, but a new order from which the classical Cartesian order could be abstracted in some suitable limit. This structure, we believed, would provide a more natural way of accounting for quantum phenomena.

The basic ideas of "structure" had already been introduced by Eddington (1958) when he raised the question: "What sort of thing is it that I know"? For him the answer was *structure*, structure that could be captured by mathematics. For example, the concept of space is not an empty "container", but a relationship of the ensemble of movements that is experienced as we probe our surroundings, using light signals or other suitable physical processes. For Eddington, the structure of these experiences could be captured by a group, which in the relativistic case would be the Lorentz group, giving rise to Minkowski space-time. Of course in the presence of a gravitational field, this group must be replaced by a larger group, the group of general coordinate transformations but for Penrose the conformal group was general enough to be explored initially. When we come to quantum phenomena, Weyl (1931) pointed out that we must turn our attention not to the group, but to the group algebra.

## 1.3 The Role of Clifford Algebras

However, again, I go too fast because initially Bohm and I thought that a natural mathematical expression of this structure would be provided by combinatorial topology alone (Bohm et al. 1970). Although this provided some interesting insights, it misses a vital ingredient, namely, the activity or movement that was necessary to describe process. But then I noticed that Penrose's spin network had Clifford algebras at its heart, the algebra that Onsager used to solve the two-dimensional Ising model. Could it be that the combinatorial aspects could be captured by an algebra itself, so that we could use algebras to describe a dynamic structure-process?

To my surprise I found that Clifford (1882) was led to his algebra not by thinking of a quantum system, but by considering the dynamical activity of classical mechanical systems. He noticed that Hamilton's quaternion algebra, a way of describing rotations in space through action, could be generalized to capture the Lorentz group and even leads to the conformal group which is used in twistor theory. Algebraic elements could be understood in terms of how movements could be combined to form new movements. Clifford introduced terms like "versors", "rotators", and "motors" emphasizing activity. Unfortunately these ideas seemed to add nothing new to physics that was not already described more simply by the vector calculus, so the algebraic approach was ignored. However, that changed when Dirac, faced with the negative energies appearing in the relativistic generalization of the Schrödinger equation, rediscovered the Clifford algebra. It provided a description of spin, relativity and the twistor in one algebraic hierarchy.

Unfortunately the appearance of the Dirac-Clifford algebra did not lead to a reconsideration of Clifford's ideas. Rather the algebra was seen as a generalization of the quantum operator algebra that was already used in the standard Hilbert space formalism taught to undergraduates. In that approach the wave function played a key role and gave rise to the so-called "wave-particle duality", a notion that I find very unhelpful, being a totally confused idea. Somehow this wave function is used to describe the so-called "state of the system" which was, in turn, assumed to evolve in the Cartesian order of space and time. While this approach was a predictive success, it has many, as yet, unsolved interpretational problems, such as the measurement problem, schizophrenic cats and the like. All of these could be handled as a set of rules for getting "correct" results, but one is left with the uneasy feeling that something is not quite right because the nature of the physical processes themselves remains very unclear.

This view was shared by Hans Primas who posed the question: Why a Hilbert space model? He then explained that Hilbert space was but a particular representation of a more general quantum mechanics. The algebra emphasizes a non-commutative structure, a structure that has its origins in the early work of Born and Jordan (1925). For Primas and Müller-Herold (1978)[1]

---

[1] Note that what they called a B*-algebra in 1978 is nowadays usually referred to as a C*-algebra.

algebraic quantum mechanics starts with an abstract B\*-algebra, $\mathcal{A}$, of observables. From this algebraic realization of quantum mechanics, we can get the corresponding Hilbert-space model $\mathcal{H}$... as the universal representation $(\pi, \mathcal{H})$ of the B\*-algebra $\mathcal{A}$.

Thus Hilbert space is a mere representation, but a representation of what? Could algebraic structure itself provide a description of structure-process and in doing so, clarify the nature of quantum processes?

## 2 The Propositional Calculus and Algebraic Idempotents

### 2.1 Von Neumann Algebras and a Propositional Calculus

We now come to the point where algebra meets logic. Primas highlighted the close relationship between the von Neumann algebras and orthomodular lattices of the type used in the analysis of formal logic. In fact the set of projections in a von Neumann algebra forms a complete orthomodular lattice so that investigating the properties of this lattice gives a different insight into the algebraic structure.

Projection operators are idempotents, $E^2 = E$ and because their eigenvalues are 0 and 1, they can be used to define the truth or falsity of a set of propositions. We thus have an alternative method of analysing the Schrödinger formalism in terms of a non-Boolean logic, a generalization of the Boolean logic of classical physics.

The generalized non-Boolean logic contains a new notion of *incompatible propositions*, tied intimately to the appearance of non-commuting operators. This difference led Finkelstein (1968) to conclude that the appearance of quantum processes causes a fracture in physical logic. Indeed Finkelstein showed that in this non-Boolean logic, the distributivity law of classical logic was violated.

This raises the important question as to whether this change in logic has to do with the fact that we can only obtain *incomplete knowledge* of a quantum system or whether this fact stems from a profound change in the basic *reality* underlying quantum phenomena. Bohr offered an epistemological interpretation in which he proposed that the incompatibility of propositions arises from our inability, *in principle*, to obtain complete knowledge of the system. For Bohr, quantum phenomena confirmed that there was a new principle of epistemology, namely the principle of complementarity to which all knowledge must conform. If this was a fundamental principle then, no matter what underlies appearance, it would be impossible, even in principle, to construct intuitive pictures of this underlying reality, pictures of the type used in the classical world.

However quantum phenomena occur without the need for anyone to *interpret* them or have knowledge of them. There is an actual process unfolding, independent of any observer and this fact demands an underlying ontology. As Primas and Müller-Herold (1978) insist

> ... practically all high-level theories adopt some kind of scientific realism i.e. the view that biological, chemical and physical objects have existence independent of some mind perceiving them.

The key question is then, what form this ontology is going to take. Is it going to be a "veiled reality" as suggested by d'Espagnat (2003). or do we follow Primas (1977) and insist that "the most fundamental theory has to be phrased in an *individual and ontic interpretation*? Our hope was that the notion of structure-process would provide the intuitive basis of such a fundamental theory. Any generalized theory must be based on non-commutative algebras that lie at the heart of quantum processes. Since geometry forms the basis of classical physics, its generalization, non-commutative geometry, must be the way forward to explore the nature of the underlying ontology.

Such a possibility had already been anticipated by Murray and von Neumann (1936), who presented a very detailed, but intimidating mathematical discussion of what are now called von Neumann algebras, algebras that would play a fundamental role in non-commutative geometry (Khalkhali 2009). Fortunately for the purposes of this paper we will not require this detailed knowledge as we can illustrate the essential ideas using the orthogonal Clifford algebra, a specific von Neumann algebra but one with which physicists and chemists are very familiar through the use of the Pauli $\sigma$-matrices and the Dirac $\gamma$-matrices.

What the physicist or the chemist may not realize, however, is that a Clifford algebra over a complex field is a particular example of a type $II_1$ von Neumann algebra with a Jones index of $4\cos^2(\pi/4)$ (Jones 2003). From the comments above, it should be clear that the Clifford algebra will play an important role in our discussion of a non-commutative geometry, a point of view shared by Finkelstein (1987) when he writes, "I am strongly tempted by the example of Clifford".

## 2.2 The Role of the Clifford Algebra in Non-commutative Quantum Geometry

As we have indicated, the conventional view among physicists is to regard the Clifford algebra merely as a formal mathematical device, but our introductory remarks suggest that it is more than that, describing an underlying structure-process. However, to proceed down that route means we must give up, as a fundamental form, the classical notion of a particle evolving along a well defined trajectory in an a priori given space-time. Instead we should adopt a thoroughgoing process philosophy along the lines suggested by Eddington (1958), Finkelstein (1996) and Bohm (1980).

## 2.3   What Are Quantum Particles?

To summarize then, in a process philosophy, we must give up the common sense idea that the world consists of material objects with definite size, shape and properties. But this notion has already been called into question in special relativity where we are forced to adopt a description based on the notion of a point event. There is no consistent description of an extended rigid object; a particle must be treated as a complex structure of events that can be regarded as forming a "world tube". The tube itself cannot have a sharp boundary but must be identified with a pattern of events, distinguishable, but not separate from, a complex of interrelated background events. In this approach the "particle" is a semi-stable, quasi-local feature that can preserve its form in time. However, under suitable conditions it can undergo not only quantitative changes, but also qualitative changes, in its basic elements, a phenomenon that is well-known in high energy particle physics.

In passing, note that Primas (1977) also has a similar structural notion of a "particle". He stresses that the so-called "fundamental" entities, such as electrons, protons, or quarks, must not be taken as the building blocks of reality. They are merely what he calls *patterns of reality*. For Primas these patterns emerge operationally from the empirical domain, a point to which I will return later.

A limitation of the notion of an elementary "rock-like" particle becomes even more apparent in the quantum domain. To bring the difficulty out clearly, consider the following example inspired by Weyl. Suppose we retain the classical notion of a particle with specific properties. To keep things simple, consider a quantum world in which we have a collection of objects with two distinct shapes, either spheres or cubes, and two distinct colours, either red or blue. Our task is to separate these objects into four distinct groups—red spheres, blue spheres, red cubes and blue cubes. In a classical world there is no problem, but in this quantum world, shape and color are observables, represented by non-commuting operators, their "values" being represented by their corresponding eigenvalues. This means that to separate colors and shapes, we must have two different types of observing instruments. In our case we call these instruments "spectacles".

Suppose we require to collect together an ensemble of red spheres. First we put on the "shape-distinguishing" spectacles and collect together spheres, discarding all the cubes. Then we put on the "color-distinguishing" spectacles and collect together the red spheres, discarding all the rest. We are done; we have a collection of red spheres. So what is the problem? Just recheck that the objects in the ensemble are still spheres. We use the first pair of glasses again and find that half of the objects are now cubes! No permanent *either–or* in this world. No permanent *both–and* either!

Clearly quantum phenomena do not have their existence defined in terms of classical objects with well-defined properties! Finkelstein has already stressed this feature and argues that "to speak about the wave function of the system is a syntactic error" (Finkelstein 1987, 1996). We do not simply "find" the state of a system. We have to "probe" the system with another physical process, the "observing instrument". In other words, our instruments are part of the underlying structure-process and therefore change the system itself, or better still, change the process that *is* the system. How, then, do we encompass these radically new ideas without losing features of the standard formalism that have been used with outstanding success?

Let us begin by following Eddington (1958) who suggests that the *elements of existence*, the individuals, in a process world, should be described by idempotents, $E^2 = E$. The eigenvalues, $\lambda_e$, of an idempotent are 1 or 0, existence or non-existence. In symbols

$$E^2 = E, \quad \text{with} \quad \lambda_e = 1 \text{ or } 0.$$

If all idempotents commute, as in classical physics, existence is always well defined. We have a Boolean logic. In quantum theory we have a difference, idempotents do not always commute

$$[E_a, E_b] \neq 0.$$

What then of existence?

$$\text{Either } E_a \text{ or } E_b, \quad \text{never both } E_a \text{ and } E_b.$$

Existence, non-existence and in between? This is the consequence of a non-Boolean logic.

## 2.4   Idempotents and Clifford Algebras

The suggestion is that the idempotent will provide a means of focusing on the *sub-process* that *is* the individual. The individual is a process that is continually changing into itself, $E \cdot E = E$. While probing the individual, the process may change the quality of the idempotent, it nevertheless remains an idempotent, enabling us to track the individual as a sub-process within the whole structure-process. In an algebra, an idempotent can be used to define a set of elements within a minimal left ideal of the total algebra. These elements carry all the information contained in the "wave" function but now have the advantage of being an integral part of the whole algebra.

In a semi-simple algebra, we can always form an element of such an ideal by writing $\Psi_L(\mathcal{A}) = \psi_L(\mathcal{A})E$. Mathematically we are constructing a left module or left vector space, but we need not be familiar with this mathematical structure to see how it works. Consider a spin-half system which requires the observables to be expressed in terms of the Pauli spin matrices. As is well known the spin "wave" function is a column two-matrix, the spinor,

$$\Psi = \begin{pmatrix} \psi_1 \\ \psi_2 \end{pmatrix}.$$

From the algebraic point of view, the Pauli spin matrices define the Clifford algebra $C_{3,0}(\sigma)$ generated by the three Pauli spin matrices $\sigma_i$. An element of a minimal left ideal can be written in the form $\Psi_L(\sigma) = \psi_L(\sigma)E$ where $E$ is some idempotent. It is conventional to choose $E = (1 + \sigma_3)/2$, which breaks the rotational symmetry and defines a preferred $z$-axis while $\psi_L(\sigma) \in \mathcal{A}$.

If we then polar decompose the algebraic spinor, we can write $\Psi_L(\sigma) = RU$ where $U = U^{\dagger}$ and $R$ is a positive definite matrix. It is then easy to show that the spinor can be written in the form

$$\Psi_L(\sigma) = g_0 + g_1\sigma_{23} + g_2\sigma_{13} + g_3\sigma_{12}, \qquad g_i \in \mathbb{R}.$$

Here we have written the elements of the algebra in terms of Pauli matrices, $\sigma_{ij} = \sigma_i\sigma_j$, a rotor. To make contact with the usual spinor, we have the identities

$$g_0 = (\psi_1^* + \psi_1)/2, \qquad g_1 = i(\psi_2^* - \psi_2)/2,$$
$$g_2 = (\psi_2^* + \psi_2)/2, \qquad g_3 = i(\psi_1^* - \psi_1)/2. \tag{1}$$

Let us emphasize again that we have chosen the specific idempotent $E = (1 + \sigma_3)/2$, which means that we have broken the spherical symmetry by picking a specific direction, conventionally the $z$-axis. This is usually done by introducing a homogeneous magnetic field, so the choice of idempotent is defined *operationally*, just as Primas' patterns are defined operationally. In other words we are changing the process that *is* the system under investigation. In Wheeler's words (Wheeler 1991, p. 286), we are participating in the process to induce a change in the process that constitutes the system.

This is exactly what we need to account for our toy model of a quantum world using "shapes" and "colors". The change that we find when checking the content of the final ensemble arises from the participatory nature of our "instrument". Looking through the "quantum spectacles" is not a passive process, it is an *action*, which must not be thought of as a mere "disturbance". It is an inescapable change in the structure-process that *is* the system. More details of this idea will be found in Hiley and Frescura (1980) and in Hiley and Callaghan (2010).

This example explains very succinctly how the Pauli algebraic spinor appears and is used in the description of the algebra. It is easily generalized to the Dirac spinor and indeed the twistor, which is a semi-spinor of the conformal Clifford algebra. These Clifford algebras form a hierarchy or tower of algebras, $C_{3,0} \to C_{1,3} \to C_{4,1} \to C_{2,4}$ of the type considered by Jones (1986). It is interesting to note that the Schrödinger "wave" function can also be considered as an element of a minimal left ideal in the Clifford algebra $C_{0,1}$, with the quaternions appearing in $C_{0,2}$.

In addition to elements of the left ideal, we also have dual elements, $\Psi_R(\mathcal{A}) = E\psi_R(\mathcal{A})$, chosen from an appropriate minimal right ideal. This enables us to give a complete specification of the structure-process of an individual system by writing

$$\rho_c(\mathcal{A}) = \Psi_L(\mathcal{A})\widetilde{\Psi}_L(\mathcal{A})$$

where $\rho_c(\mathcal{A})$ is an element that characterizes the system. It is the algebraic analog of the density matrix.

If we define $\widetilde{\Psi}_L(\mathcal{A}) = \Psi_R(\mathcal{A}) = E\widetilde{\psi}_L(\mathcal{A})$ then, by a suitable choice of the tilde operation, we find $\rho_c^2 = \rho_c$, a signature of what is known in the standard approach as a *pure state*. It should be noted that the corresponding dual element introduced by Primas and M?ller-Herold was called a *normalized positive linear functional*. Using this additional mathematical structure, we have the possibility of a generalization to mixed states, but in this paper we confine our attention to pure states for simplicity.

As well as rotational symmetries, we must also consider translational symmetries, which implies turning our attention to the Heisenberg algebra. Here there is a technical problem because this algebra is nilpotent and therefore does not contain any idempotents. However, Schönberg (1957), and later Hiley (2001), showed that it was possible to extend this algebra by adding sets of idempotents to form a symplectic Clifford algebra (Crumeyrolle 1990). This then enables us to employ similar techniques to those used in the orthogonal Clifford algebra. One is then able to find time evolution equations that correspond to the Heisenberg equations of motion.

The characteristic element $\rho_c(\mathcal{A})$ can now be subjected to both left and right translations to determine two fundamental time evolution equations,

$$i[(\partial_t \Phi_L)\widetilde{\Phi}_L + \Phi_L(\partial_t \widetilde{\Phi}_L)] = i\partial_t \rho_c = (\overrightarrow{H} \Phi_L)\widetilde{\Phi}_L - \Phi_L(\widetilde{\Phi}_L \overleftarrow{H}) \qquad (2)$$

and

$$i[(\partial_t \Phi_L)\widetilde{\Phi}_L - \Phi_L(\partial_t \widetilde{\Phi}_L)] = (\overrightarrow{H} \Phi_L)\widetilde{\Phi}_L + \Phi_L(\widetilde{\Phi}_L \overleftarrow{H}). \qquad (3)$$

We now have the possibility of two forms of Hamiltonian $\overrightarrow{H} = \overrightarrow{H}(\overrightarrow{D}, V, m)$ and $\overleftarrow{H} = \overleftarrow{H}(\overleftarrow{D}, V, m)$, emphasizing the distinction between left and right translations. We will not derive these equations here (see Hiley and Callaghan 2010); nevertheless we will use them in the next section. We merely note that Eq. (2) is the quantum Liouville equation expressing the conservation of probability, while Eq. (3) is the quantum Hamilton-Jacobi equation expressing the conservation of energy. A detailed discussion of these equations will be found in Hiley (2015).

## 3    The Implicate and Explicate Order

We must now return to discuss the relation between the non-Boolean structure and its Boolean substructures. Primas (1977) offers a formal way to understand the relationship between these two logics in terms of a specific physical process. We will explain his position in the following way.

We have argued that there is no such thing as a direct, faithful observation in a quantum process. However, as Bohr has pointed out, the results of any observation must be unambiguously described in terms of a Boolean structure. This is the only way we can unambiguously communicate the results of an experiment. How then do we understand the Boolean aspects of a fundamentally non-Boolean process?

Primas suggests that the results of an experiment can be understood as a *pattern* that is formed by detaching ourselves, and our instruments, from properties that we consider to be non-essential. He calls the total process, the *factual* domain $\mathcal{F}_\alpha$, which he distinguishes from the empirical domain $\mathcal{E}_\alpha$ defined *operationally* as the result of the $\alpha$th pattern recognition technique. The factual domain is non-Boolean and a-local, while the empirical domain is a Boolean and local structure. The link between theory and experiment is then regarded as a mapping $\mathcal{F}_\alpha \to \mathcal{E}_\alpha$ which is not required to be one-to-one.

Bohm (1980) has made, in essence, a similar proposal to understand the relation between Boolean and non-Boolean aspects of physical processes, but in terms of a more general language. Structure-process is defined in terms of an algebra in which the individual elements of the algebra, like words, take their implicit meaning from the way in which the algebra as a whole is used. For example the symbols in the Pauli Clifford algebra take their meaning from the rotational symmetries we experience as we rotate in space.

In such a structure, all the spin components cannot be made explicit by the same action. The spin in the $z$-direction can be made explicit, while the other components remain implicit. More generally, as is well known, an ensemble of properties corresponding to mutually commuting observables can be made explicit together. This subset of elements forms a Boolean substructure within the more general non-Boolean structure. Bohm called these substructures explicate orders, while the total non-Boolean structure was called the implicate order.

I have used examples from gestalt psychology as a metaphor to illustrate the notions of the implicate and explicate order. The young lady/old lady gestalt illustrates succinctly what is involved. Our perception constructs or "explicates" a Boolean pattern, say the young lady, by ignoring some of the details in the drawing. When none of the details are ignored, we have a non-Boolean structure. However, metaphors are limited and a deeper analysis based on Eq. (3) shows that a projection actually creates the explicate order. It creates a Boolean substructure within the non-Boolean totality.

To see how the projection comes in, let us write Eqs. (2) and (3) in a more familiar notation,

$$i\partial\rho = (H|\phi\rangle)\langle\phi| - |\phi\rangle(\langle\phi|H) \tag{4}$$

and

$$i[(\partial_t|\phi\rangle)\langle\phi| - |\phi\rangle(\partial_t\langle\phi|)] = (H|\phi\rangle)\langle\phi| + |\phi\rangle(\langle\phi|H). \tag{5}$$

Now introduce the projection operator $P_a = |a\rangle\langle a|$ and take the trace so that Eq. (4) becomes

$$\frac{\partial P(a)}{\partial t} + \langle [\rho_c, H]_- \rangle_a = 0 \tag{6}$$

while Eq. (5) becomes

$$2P(a)\frac{\partial S_a}{\partial t} + \langle [\rho_c, H]_+ \rangle_a = 0. \tag{7}$$

To bring out what this means, let us consider an harmonic oscillator Hamiltonian $\hat{H} = \hat{p}^2/2m + K\hat{x}^2/2$ and choose the projection operator $P_x = |x\rangle\langle x|$ so that Eq. (6) becomes

$$\frac{\partial P_x}{\partial t} + \nabla_x \cdot \left( P_x \frac{\nabla_x S_x}{m} \right) = 0.$$

This is just the equation for the conservation of probability in position space. Using the same procedure on Eq. (7) finally gives us

$$\frac{\partial S_x}{\partial t} + \frac{1}{2m} \left( \frac{\partial S_x}{\partial x} \right)^2 - \frac{1}{2mR_x} \left( \frac{\partial^2 R_x}{\partial x^2} \right) + \frac{Kx^2}{2} = 0$$

which is just the quantum Hamilton-Jacobi equation for the harmonic oscillator. This is simply the equation Bohm obtained by taking the real part of the Schrödinger equation under polar decomposition of the wave function. This equation contains the quantum potential

$$Q = -\frac{1}{2mR_x} \left( \frac{\partial^2 R_x}{\partial x^2} \right). \tag{8}$$

Notice that this potential does not appear in the algebraic equation (3) which we are regarding as a description of the implicate order. It only appears in the projected space. This space is a Boolean phase space constructed with $(x, p_B(x))$ where $p_B(x)$ is the Bohm or local momentum. It is in this phase space that trajectories have been constructed by Philippidis et al. (1979). Thus we have constructed a Boolean explicate order.

We could choose another projection operator $P_p = |p\rangle\langle p|$ so that the two Eqs. (2) and (3) now become

$$\frac{\partial P_p}{\partial t} + \nabla_p \cdot \left( P_p \frac{\nabla_p S_p}{m} \right) = 0$$

and

$$\frac{\partial S_p}{\partial t} + \frac{p^2}{2m} + \frac{K}{2} \left(\frac{\partial S_p}{\partial p}\right)^2 - \frac{K}{2R_p} \left(\frac{\partial^2 R_p}{\partial p^2}\right) = 0.$$

This enables us to project out another Boolean phase space based, this time, on $(x_B(p), p)$ where $x_B(p) = -\left(\frac{\partial S_p}{\partial p}\right)$. Thus, using the momentum representation we have constructed another explicate order and thereby revealed $x, p$ symmetry—a symmetry that Heisenberg (1958, p. 118) claimed was *not* present in the Bohm approach.

Bohm chose the $x$-representation as a preferred representation simply because he saw a problem in representing the Coulomb potential in the $p$-representation. However, for other potentials there is no difficulty. Indeed Brown and Hiley (2000) showed how the approach worked in the particular case of a cubic potential.

Another criticism that is often made of the Bohm approach is that it does not work for the relativistic Dirac particle. However Hiley and Callaghan (2012) have shown that we can obtain Lorentz invariant analogs of Eqs. (2) and (3) which can then be put into the form of a relativistic quantum Hamilton-Jacobi equation. To do this we need to use the orthogonal Clifford algebra $C_{1,3}$. The expression for the quantum potential is more complicated but can be shown to reduce to the expression (8) in the non-relativistic limit (Hiley and Callaghan 2012).

These examples show what is involved in what Primas calls *pattern recognition*. It is not a "passive" recognition, it actually involves an *active construction* of the Boolean pattern. But in doing so new features can be introduced, as Primas points out. In the case of the Boolean phase space considered above, it is the appearance of the quantum potential which can be considered as the appearance of a force.

This is not unlike the nature of the gravitational force which only appears when we project the curved space-time geodesic to a flat Minkowski space-time. However there is a significant difference in that the curvature of space-time is universal, whereas the quantum potential is, in a sense, "private", being shared by a group of *entangled* particles. We could have a situation arising where the quantum potential of one group of entangled particles can be very different from the quantum potential of another entangled group if the groups are non-interacting but nevertheless share the same region of space-time. The groups do not experience a common quantum potential, it is not universal since they only experience the quantum potential of their own group.

## 4 Conclusion

In this paper I have given a brief view of a new way of looking at quantum phenomena that Hans Primas was one of the first to draw to our attention. His and my approaches did not develop in parallel after his pioneering contributions addressed in this paper—other articles in this volume focus on his use of later developments of non-commutative mathematics after the 1970s. However, I will always be grateful to Hans for his early work and our subsequent discussions which, although at times heated, always provided new insights.

**Acknowledgments**  I should like to thank Glen Dennis for his suggestions and helpful comments.

# References

Abramsky, S., and Coecke, B. (2004): A categorical semantics of quantum protocols. In *Logic in Computer Science*, IEEE Computer Society, Washington DC, pp. 415–425.

Bohm, D. (1965): Space, time, and the quantum theory understood in terms of discrete structural process. *Proceedings of the International Conference on Elementary Particles*, Kyoto, pp. 252–287.

Bohm, D. (1971): Space-time geometry as an abstraction from spinor ordering. In *Perspectives in Quantum Theory: Essays in Honour of Alfred Landé*, ed. by W. Yourgrau, MIT Press, Cambridge, pp. 78–90.

Bohm, D. (1980): *Wholeness and the Implicate Order*, Routledge, London.

Bohm, D.J., Hiley, B.J., and Stuart, A.E.G. (1970): On a new mode of description in physics. *International Journal of Theoretical Physics* **3**, 171–183.

Born M., and Jordan P. (1925): Zur Quantenmechanik. *Zeitschrift für Physik* **34**, 858–888.

Brown, M.R., and Hiley, B.J. (2000): Schrödinger revisited: the role of Dirac's "standard" ket in the algebraic approach. Preprint accessible at at http://arxiv.org/abs/quant-ph/0005026.

Clifford W.K. (1882): Further note on biquaternions. In *Mathematical Papers XLII*, ed. by R. Tucker, Macmillan, London, pp. 385–394.

Coecke, B. (2005): Kindergarten quantum mechanics. In *Quantum Theory: Reconsiderations of Foundations III*, ed. by A. Khrennikov, AIP Press, New York, pp. 81–98.

Crumeyrolle A. (1990): *Orthogonal and Symplectic Clifford Algebras: Spinor Structures*, Kluwer, Dordrecht.

Domb, C., and Hiley, B.J. (1962): On the method of Yvon in crystal statistics. *Proceedings of the Royal Society* **A268**, 506–526.

Eddington, A.S. (1936): *Relativity Theory of Protons and Electrons*, Cambridge University Press, Cambridge.

Eddington, A.S. (1958): *The Philosophy of Physical Science*, University of Michigan Press, Ann Arbor.

Finkelstein, D. (1968): Matter, space and logic. In *Boston Studies in the Philosophy of Science V*, ed. by R.S. Cohen and M.W. Wartowsky, Reidel, Dordrecht, pp. 199–215.

d'Espagnat, B. (2003): *Veiled Reality: An Analysis of Present-Day Quantum Mechanical Concepts*, Westview Press, Boulder.

Finkelstein, D. (1969): Matter, space, and logic. In *Boston Studies in the Philosophy of Science V*, ed. by R.S. Cohen and M.W. Wartowsky, Reidel, Dordrecht, pp. 199–215.

Finkelstein, D. (1987): All is flux. In *Quantum Implications: Essays in Honour of David Bohm*, ed. by B.J. Hiley and D. Peat, D., Routledge and Kegan Paul, London, pp. 289–294.

Finkelstein, D.R. (1996): *Quantum Relativity: A Synthesis of the Ideas of Einstein and Heisenberg*, Springer, Berlin.

Grassmann, H.G. (1894): *Gesammelte mathematische und physikalische Werke*, Teubner, Leipzig.

Grassmann, H.G. (1995): *A New Branch of Mathematics: the Ausdehnungslehre of 1844 and Other Works*, translated by L.C. Kannenberg, Open Court, Chicago.

Hamilton, W. R. (1967): *Mathematical Papers, Vol. 3: Algebra*, Cambridge University Press, Cambridge.

Heisenberg, W. (1958): *Physics and Philosophy: The Revolution in Modern Science*, George Allen and Unwin, London.

Hiley, B.J. (2001): A note on the role of idempotents in the extended Heisenberg Algebra. In *Implications (ANPA 22)*, Alternative Natural Philosophy Association, Cambridge, pp. 107–121.

Hiley, B.J. (2015): On the relationship between the Moyal algebra and the quantum operator algebra of von Neumann. *Journal of Computational Electronics* **14**, 869–878.

Hiley, B.J., Burke, T., and Finney, J. (1977): Self-avoiding walks on irregular structures. *Journal of Physics* **A10**, 197–204.

Hiley, B.J., and Callaghan, R.E. (2010): The Clifford algebra approach to quantum mechanics A: The Schrödinger and Pauli particles. Preprint accessible at arXiv:1011.4031.

Hiley, B.J., and Callaghan, R.E. (2012): Clifford algebras and the Dirac-Bohm quantum Hamilton-Jacobi equation. *Foundations of Physics* **42**, 192–208.

Hiley, B.J., and Frescura, F.A.M. (1980): The implicate order, algebras and the spinor. *Foundations of Physics* **10**, 7–31.

Jones, V.F.R. (1986): A new knot polynomial and von Neumann algebras. *Notices of the American Mathematical Society* **33**, 219–225.

Jones, V.F.R. (2003): Von Neumann algebras. Lecture notes accessible at http://www.math.berkeley.edu/vfr/MATH20909/VonNeumann2009.pdf.

Kauffman, L.H. (2001): *Knots and Physics*, World Scientific, Singapore.

Khalkhali, M. (2009): *Basic Non-Commutative Geometry*, EMS Publishing, Zurich.

Murray, F.J., and von Neumann, J. (1936): On rings of operators. *Annals of Mathematics* **37**, 116–229.

Onsager, L. (1944): Crystal statistics. I. A two-dimensional model with an order-disorder transition. *Physical Review* **65**, 117–149.

Penrose, R. (1967): Twistor algebra. *Journal of Mathematical Physics* **8**, 345–366.

Penrose, R. (1971): Angular momentum: A combinatorial approach to space-time. In *Quantum Theory and Beyond*, ed. by T. Bastin, Cambridge University Press, Cambridge, pp. 151–180.

Philippidis, C., Dewdney, C., and Hiley, B.J. (1979): Quantum interference and the quantum potential. *Nuovo Cimento* **52B**, 15–28.

Primas, H.(1977): Theory reduction and non-Boolean theories. *Journal of Mathematical Biology* **4**, 281–301.

Primas, H., and Müller-Herold, U. (1978): Quantum mechanical system theory: A unifying framework for observations and stochastic processes in quantum mechanics. *Advances in Chemical Physics* **38**, 1–107.

Schönberg, M. (1957): Quantum mechanics and geometry. *Anais da Academia Brasileira de Ciencias* **29**, 473–485.

Weyl, H. (1931): *The Theory of Groups and Quantum Mechanics*, Dover, London.

Wheeler, J.A. (1991): *At Home in the Universe*, AIP Press, New York.

# Non-commutative Structures from Quantum Physics to Consciousness Studies

Harald Atmanspacher

**Abstract** It has been an old idea by Niels Bohr, one of the architects of quantum physics, that central features of quantum theory, such as complementarity, are also of pivotal significance beyond the domain of physics. But Bohr—and others, such as Wolfgang Pauli—never elaborated this idea in concrete detail, and for a long time no one else did so either. This situation has changed: there are now a number of research programs applying key notions of quantum theory in areas of knowledge outside physics. In his typical way, both insurgent and conservative, Hans Primas has critically supported and crucially contributed to these developments. There are two major extraphysical directions in which non-commuting operations, the basis of complementarity, have been applied in the past 20 years. One of them refers to fertile new insights in psychology and cognitive science, due to which non-commutativity is a core feature of various kinds of decision-making processes. Meanwhile, there is a number of research groups worldwide who study these and other cognitive processes using quantum concepts. The other direction is closely related to a topic that interested Primas since his student days: the philosophical conjecture, developed by Pauli and C.G. Jung, that the mental and the physical are complementary aspects of one underlying reality that itself is psychophysically neutral. In his most recent work, Primas exploited this framework to explore the relation between mental and physical time.

## 1 Introduction

Hans Primas was already in his early 60s when I came into closer contact with him, at the Cortona Week in 1991 that was later turned into an integral part of the curriculum of the Swiss Federal Institute of Technology (ETH) at Zurich. The topic of the week was "Metamorphoses", Primas was one of the keynote speakers, and I attended the conference as a postdoc at the Max-Planck Institute for Extraterrestrial Physics at Garching.

H. Atmanspacher (✉)
Collegium Helveticum, University and ETH Zurich, Zurich, Switzerland
e-mail: atmanspacher@collegium.ethz.ch

© Springer International Publishing Switzerland 2016
H. Atmanspacher and U. Müller-Herold (eds.), *From Chemistry to Consciousness*, DOI 10.1007/978-3-319-43573-2_8

127

At the time, I was working on a publication about Wolfgang Pauli and alchemy, a theme that fascinated me after I had learned that the series of dreams Jung reports in his *Psychology and Alchemy* actually was from Pauli's dream diary. Since I knew that Primas was interested in the exchange between Pauli and Jung, I had sent him a preprint of the first part of my work some time before the Cortona meeting, but hadn't heard anything back from him. When we spotted each other in front of the lecture hall, we immediately started discussing—as if this wasn't the first time we met. For me, a young scientist, it was especially impressive how a scholar with his accomplishments and worldwide reputation had acquired the inner freedom to attend to themes that many other scientists would readily dismiss as abstruse (or worse).

Over the days, our conversation expanded from the Pauli-Jung dialog in particular to more general questions, all related to the age-old topic of the relationship between the mental and the physical.[1] I will get back to this in Sect. 4, which addresses the philosophical framework for mind-matter relations, a chapter in speculative metaphysics, that we could later reconstruct from more or less scattered remarks in articles by Pauli and Jung as well as from their correspondence.

Yet the conversations with Primas at Cortona had an additional side which I could not possibly have anticipated. In the late 1980s, he had started to work on how various interpretations of quantum physics might be intelligible within the formal framework of algebraic quantum theory. This became a topic of discussion almost the first day of the conference—actually I should say the first night, when I found myself embroiled in his explanations of C*-and W*-algebras, GNS-constructions, KMS-states and so on at the bar of the Cortona hotel. Obviously, my ignorance sparked his teaching instincts, and so we spent almost every night with a high-density crash course on algebraic quantum theory and the way it helps understanding a number of conceptual riddles of quantum physics—spiced with one or another drink from the bar.

For me this was a revelation. Many of the issues that are hardly mentioned and even less explained in the regular quantum mechanics courses became transparent and fell into place. A subject that I had learned to accept as both formally demanding and conceptually counterintuitive was transformed into a coherent framework of old puzzles appearing in new and consistent connections. Needless to say, my acquaintance with the algebraic approach, as rewarding as I experienced it at first sight, required much more work in detail to become a solid basis for thinking—not to mention truly original work in the field, for which I am not knowledgable enough until today.

At a moderate level, though, the algebraic approach became familiar to me to an extent that made it possible for Primas and myself to discuss and, later, publish our ideas together with their implications for the philosophy of physics. A basic result of this work was the insight that many of the alleged mysteries of quantum theory originated in two basic classes of category mistakes: one of them arising from classically misguided discussions of quantum phenomena, the other from the

---

[1]Needless to say, this became the focus of Primas' interests way after his early work on physical and theoretical chemistry, which is addressed in the chapters by Ernst, Bodenhausen, and Müller-Herold in this volume.

confusion of ontic and epistemic descriptions of quantum systems. This second point will be addressed in Sect. 2.

Eventually, there was one more significant step that I became infected with through our interactions: the mathematical tools that algebraic quantum theory uses are not necessarily restricted to physics. The non-commutativity of operations is at the heart of these tools, and Primas has been a source of inspiration and encouragement to try and apply them to areas beyond the limits of physics. This novel field of research, much of which concerns topics of psychology and cognitive science, has spread out to numerous places across the globe by now, with considerable initial success and with a lot of momentum to expand, as will be discussed in Sect. 3.

Hans Primas in his office at floor G in the ETH chemistry building at Universitätsstrasse Zurich in the mid 1990s: "I am not Boolean."

As Primas showed in his *opus magnum* of 1981, non-commutative operations in physics are isomorphic to non-Boolean lattices of propositions (about such operations) in logic. In a nutshell, this logic entails that binary yes-no alternatives are too limited to understand our minds and the world around us. Non-Boolean logic rejects the law of the excluded middle, the *tertium non datur*. It expresses the fact that we need more for truth judgments than the categories of right and wrong, and that the context of a statement is often decisive for its significance. In discussion with Primas one could occasionally experience that what he said one day seemed to contradict what he said another day. Upon requests for clarification, it happened more than once that he mastered this challenge with the sibylline remark: "I am not Boolean".

## 2   Ontic and Epistemic Descriptions

### 2.1   Kinds of Realism

Most working scientists believe that there is an external world, which has the status of a reality to be explored by science. The goal of science is to achieve knowledge about how this external world is constituted and develops. Although scientific methodology requires observations and measurements for this purpose, the reality to be described is believed to "exist" independent of its possible empirical accessibility. This view is succinctly formulated by Einstein (1949a, p. 81):

> Physics is an attempt conceptually to grasp reality as it is thought independently of its being observed.

On the other hand, there is a different stance to the effect that quantum theory does not admit such an observation-independent realism. This view, which has been perpetuated in many modern monographs and textbooks, goes back to Bohr's claim that in quantum theory a realism with respect to measuring instruments is the only possible realism (sometimes even referred to as "anti-realism"). According to Bohr (quoted in Petersen 1963),

> it is wrong to think that the task of physics is to find out how nature is. Physics concerns what we can say about nature.

The two quotations by Einstein and Bohr indicate a basic point of disagreement between the two in their ongoing conversations concerning the interpretation of quantum mechanics in the 1920s and 1930s (compare Bohr 1949 and Einstein 1949b). Bohr focused on what we could know about and infer from observed quantum phenomena. By contrast, Einstein's position led him to consider Bohr's characterization of quantum theory as incomplete.

Now it would be premature to infer from Einstein's realist position that observations of features of the assumed observer-independent reality exhibit that reality as an exact image, by a one-to-one mapping as it were. And it would be equally premature to think that Bohr denied that there is a world out there. His stance only insists that all we can know about it is restricted to be relative to observations and the way we talk about them. So the contrast between the two positions may ultimately be less sharp than the two quotes might indicate.

Many discussions about realism in science nevertheless took the positions by Bohr and Einstein as a blueprint for the belief that arguing in favor of one of them implies logically to argue against he other. Primas realized early on that this strictly Boolean move might be mistaken. In order to introduce a more advanced position, the first thing he did was to look for a way in which the differences between them can be formalized explicitly and in detail. In the late 1980s, he discovered that algebraic quantum theory offers exactly such an option, which can be combined with the philosophical distinction of ontic and epistemic descriptions as introduced by Scheibe (1964, 1973).

## 2.2 Ontic and Epistemic States and Observables

In a series of papers starting in 1990, Primas picked up this philosophically grounded distinction and connected ontic and epistemic perspectives to particular elements of the algebraic approach to quantum theory. Some relevant articles are Primas (1990, 1991, 1993), Amann and Primas (1997), Amann and Atmanspacher (1999), Atmanspacher and Primas (2003).[2] It should be noted that these papers are essentially restricted to a Galilei-invariant version of quantum theory, leaving aside its extension toward relativistic frameworks.

*Ontic states* encode all properties of a system exhaustively: An ontic state is "just the way it is", without reference to epistemic knowledge or ignorance (due to observation or measurement). Ontic states are the referents of descriptions of *individual systems*, represented pointwise in an appropriate state space. The properties of the system are treated as *intrinsic properties*, as context-free as possible. Insofar as ontic states are observation-independent, the associated intrinsic properties are *empirically inaccessible*. They are *idealizations*, which is expressed by the fact that they refer to closed systems with a unitary (reversible) dynamics.

*Epistemic states* encode our (usually non-exhaustive) knowledge of the properties of a system, based on a discrete partition of the relevant state space. The referents of *statistical descriptions* are epistemic states (ensembles with probability distributions). The properties of the system are treated as *contextual properties*, i.e. they are defined with respect to a particular context to be chosen. Contextual properties associated with epistemic states are *empirically accessible* by observation and measurement. They refer to the realistic situation of open systems, which are governed by a semigroup (irreversible) dynamics.

The proposal that Primas made was essentially a mapping of intrinsic properties to elements of a C*-algebra $\mathcal{A}$ of observables, whereas contextual properties are mapped onto elements of a W*-algebra $\mathcal{M}$ of observables. The dual $\mathcal{A}^*$ of $\mathcal{A}$ is then the space of ontic states, whereas the predual $\mathcal{M}_*$ of $\mathcal{M}$ is the space of epistemic states.[3] A particular feature of quantum systems is that they posses observables that do not commute (see also Sect. 3.1). If a system has only non-commuting observables, it is called a *factor*. If a system has both commuting and non-commuting observables, the commuting observables (also referred to as classical observables) are elements of the so-called *center* of the algebra.

---

[2] For a while, Primas explored a different terminology, calling ontic descriptions "endo-descriptions" and epistemic descriptions "exo-descriptions" (Primas 1994a). In this terminology "endo" was meant to indicate a perspective "from within", without external tools of observation, and "exo" was meant to indicate that a system is addressed "from outside", as coupled to an environment, including observational tools. The endo-exo distinction did not prevail, however, and he returned to the ontic-epistemic terminology later on.

[3] Note that W*-algebras are also called von Neumann algebras. The term C*-algebra replaces the old notion of a B*-algebra, which is not used any more today. See pertinent textbooks for further details, which exceed the scope of this article.

## 2.3   Measurement

Given that a major conceptual difference between ontic and epistemic states is the issue of empirical access, a crucial feature of the relation between ontic and epistemic states is the transition from unobserved to observed states. In the literature on quantum theory, this transition is addressed by the notion of measurement, and the problem of how to describe it properly. In Primas' terms, this can be rephrased by the question of how contextual properties can be constructed from intrinsic properties. In more formal terms, the concept of measurement is tightly connected to the way in which a (representation-free) C*-algebra is connected to its representation by a W*-algebra (for instance a Hilbert space representation).

The algebraic framework offers such a representation, known as the GNS-representation, according to Gel'fand, Neimark, and Segal. Skipping the formal details, choosing a context and implementing it in $\mathcal{A}^*$, the space of ontic states, generates a contextual topology (coarser than that of $\mathcal{A}^*$) with equivalence classes of states. The properties associated with those equivalence classes are the contextual properties determined by the deliberately chosen context. This context is usually not prescribed at the C*-level of $\mathcal{A}$. In contrast to the Stone-von Neumann theorem, stating that all representations of a *finite* C*-system are unitarily equivalent, the general situation of *infinitely many* degrees of freedom leads to W*-representations that are inequivalent.

Primas often insisted that a number of popular approaches to the measurement problem are ill-defined, non-viable, or even absurd. Key requirements that he saw for a reasonable account are that a measurement process takes time (i.e., is not instantaneous) and must be considered a real process (i.e., not merely a projection onto subspaces of a Hilbert space). Moreover, acts of measurement must produce disjoint states (compare the contribution by Giulini in this volume), and the measurement outcome must be described as a classical, irreversible fact (that cannot be undone).

In this spirit he advocated, most expressively in Primas (1997), an approach based on a dynamical spin chain model originally suggested by Hepp (1972) and refined by Lockhart and Misra (1986). In this approach, classicality emerges gradually as a function of time, which is formally achieved by representing measurement as a K-flow of a W*-system within a statistical, epistemic description.[4]

---

[4]Note that such a description disregards the conceptual point that the unmeasured state of a system, which is transformed into a measured state through measurement, actually should be considered ontic and individual. As a reaction to this deficit, Primas (1997) wrote a manuscript on an individual description of measurement processes, which he left unpublished. A review of dynamical models of measurement, including their own proposal, was recently published by Allahverdyan et al. (2013).

## 2.4 Contextual Emergence and Relative Onticity

Primas (1994b, 1998) realized that any selected descriptive level may contain both ontic and epistemic states. This entails that a tight distinction of one fundamentally ontic and derived epistemic domains is too simplistic. However, an idea originally proposed by Quine (1969) and later utilized by Atmanspacher and Kronz (1999) comes to help here: *ontological relativity* or, in another parlance, *relative onticity*.[5]

The main motif behind this notion is to allow ontic significance for any level, from elementary particles to icecubes, bricks, and tables. One and the same descriptive framework can be construed as either ontic or epistemic, depending on which other framework it is related to. Bricks and tables will be regarded as ontic by an architect, but they will be considered highly epistemic from the perspective of a solid-state physicist. Drinks and icecubes will be regarded as ontic by a barkeeper, but they will be considered highly epistemic from the viewpoint of thermodynamics.

Quine proposed that a "most appropriate" ontology should be preferred for the interpretation of a theory, thus demanding "ontological commitment". This leaves us with the challenge of how "most appropriate" should be defined, and how corresponding descriptive frameworks are to be identified. Here is where the notion of *relevance* acquires significance. A "most appropriate" framework provides those features that are relevant for the question to be studied (cf. Atmanspacher et al. 2014). And the referents of this descriptive framework are those which Quine wants us to be ontologically committed to.

Taken seriously, this framework of thinking entails a farewell to the centuries-old conviction of an absolute fundamental ontology (usually taken as that of basic physics), to which everything else can be reduced. The corresponding move toward a *contextual emergence* (Bishop and Atmanspacher 2006) of contextual observables[6] is in strong opposition to many traditional positions in the philosophy of science until today. But in times in which fundamentalism—in science and elsewhere—appears increasingly tenuous, Quine's philosophical idea of an ontological relativity offers a viable alternative for more adequate and more balanced world views.

Coupled with an ontological commitment to context-dependent "most relevant" features in a given situation, the relativization of onticity does not mean dropping ontology altogether in favor of a postmodern salmagundi of floating beliefs. The "tyranny of relativism" (as some have called it) can be avoided by distinguishing more appropriate descriptions from less appropriate ones. The resulting picture is more subtle and more flexible than an overly bold reductive fundamentalism, and

---

[5] Similar ideas have been developed independently by van Fraassen (1980) in terms of "relevance relations", by Garfinkel (1981) in terms of "explanatory relativity", by Putnam (1981) in terms of "internal realism", and by Shimony (1993) with this "phenomenological principle". All these approaches exhibit similarities, but also differences, for instance with less, or less explicit, emphasis on issues of ontology.

[6] For further elaborations of reductive and emergence-based approaches in the philosophy of science see also the contributions by Seager and by Bishop and beim Graben in this volume.

it is more restrictive and more specific than a patchwork of arbitrarily connected opinions. Both these extremes have been frankly and frequently repudiated by Hans Primas.

# 3  Non-commuting Operations

## 3.1  Non-commutativity Within and Outside Physics

As mentioned above, a key feature of observables, e.g., $A$, $B$, in quantum theory is their *non-commutativity*, less technically referred to as *incompatibility* or *complementarity*, respectively. Its meaning is that their successive operation on objects such as a state $\psi$ of a system does not commute:

$$AB\psi \neq BA\psi.$$

An elementary example in quantum physics are spin observables with a discrete spectrum, two of which are represented by the matrices:

$$A = \begin{pmatrix} 0 & 1 \\ 1 & 0 \end{pmatrix}, \quad B = \begin{pmatrix} 1 & 0 \\ 0 & -1 \end{pmatrix}.$$

The difference of their products $AB$ and $BA$,

$$AB - BA = \begin{pmatrix} 0 & -1 \\ 1 & 0 \end{pmatrix} - \begin{pmatrix} 0 & 1 \\ -1 & 0 \end{pmatrix} = \begin{pmatrix} 0 & -2 \\ 2 & 0 \end{pmatrix},$$

does not vanish, which would be the case if the operations were commutative.

As a side remark, non-commutative algebras have an equivalent in formal logic, which leads us back to non-Boolean propositions. If an algebra contains both commuting and non-commuting elements, the corresponding logic is a partial Boolean logic. It consists of Boolean subdomains of propositions, pasted together in a globally non-Boolean fashion. As Primas (2007) argued, partial Boolean logic may still be applicable in cases where we have no clue about how to formally set up an appropriate algebra of observables.

One of the most basic operations on the state of a system is *measurement*, generally conceived as an interaction of a measuring (observing) system O with a measured (observed) system S in state $\psi$, where the measurement outcome typically is the numerical value of an observable. In systems with commuting observables, a measurement by O does not have a significant effect on S. However, in systems where observables do not commute, this effect is no longer negligible. In other words, while measurement in the commutative case simply means the registration of a value of an observable, the non-commutative case means registration of a value *plus* a change of the state $\psi$ of O. This state change is the reason why the sequence of measurement operations does make a difference.

For a long time the mathematics of non-commutative algebras has mainly, if not exclusively, been successfully used in quantum physics, the physics of systems with non-commuting observables. However, there have always been voices advocating the usage of the formalism for areas outside physics as well, starting with Niels Bohr and Wolfgang Pauli. Hans Primas belongs to the group up of those who share this vision. In one of his latest publications (Primas 2009), he states his persistent conviction that non-commutative operations and non-Boolean logic apply "far beyond quantum physics and include examples from psychology, philosophy, and engineering."

In fact, psychology and cognitive science recently saw a number of particularly convincing applications of quantum reasoning in the last two decades. This confirms the plausible assumption that non-commutative operations should be the rule rather than the exception in all kinds of mental processes. Isn't it evident that *any* observation of a mental state of a subject *always* changes that state? Here is an incomplete list of areas of research in which this basic principle has been applied (with pertinent references)[7]:

- decision and judgment processes and related paradoxes (Aerts and Aerts 1995; Pothos and Busemeyer 2009; Aerts et al. 2011);
- pattern learning and recognition on networks (Atmanspacher and Filk 2006);
- sequence effects in surveys or questionnaires (Atmanspacher and Römer 2012; Wang et al. 2014);
- bistable perception and temporal nonlocality (Atmanspacher et al. 2004, 2008; Atmanspacher and Filk 2010, 2013);
- non-separable concept combinations and semantic association (Gabora and Aerts 2002; Bruza et al. 2015).

In addition, there are other—more general—applications, neither limited to physics nor to psychology, which are worth mentioning:

- non-commutative time operators in ergodic theory and for innovation systems (Gustafson and Misra 1976; Prigogine 1980; Antoniou et al. 2016);
- non-commutative observables due to non-generating partitions in dynamical systems theory (beim Graben and Atmanspacher 2006);
- compatible and incompatible descriptions in science (Primas 1977; Prigogine 1980; Atmanspacher and beim Graben 2016)
- entanglement correlations beyond the quantum bound (Popescu and Rohrlich 1994; Dzhafarov and Kujala 2013).

---

[7]Some more commentary on the listed items can be found in Sect. 4.7 in Atmanspacher (2015). See also the monographs by Busemeyer and Bruza (2013) or by Wendt (2015).

## 3.2   Bistable Perception

One remarkably successful example is the application of non-commutative structures to the bistable perception of ambiguous stimuli, exhibiting stochastically distributed, spontaneous reversals between two possible perspectives (see Fig. 1). This example is particularly compelling because it is from psychophysics, the "most quantitative" branch of psychology, which studies the relationship between physical (external) stimuli and the perceptions they induce.

There is quite some literature trying to model features of bistable perception, which limited space does not allow me to review here. One common point in all approaches so far has been that they generically use classical modeling strategies, such as Markov models or similar. By contrast, we developed a theoretical approach which essentially decomposes the perceptual process into two kinds of dynamics that—in the spirit of quantum theory—do not commute: a reversal process $A$ and an observation process $B$, which can be plausibly represented by the matrices:

$$A = \begin{pmatrix} 0 & 1 \\ 1 & 0 \end{pmatrix}, \quad B = \begin{pmatrix} 1 & 0 \\ 0 & -1 \end{pmatrix}.$$

As we saw in Sect. 3.1, where $A$ and $B$ were introduced as spin matrices, they do not commute.

The perceptual process as a whole can then be modeled analogous to the quantum Zeno model (Misra and Sudarshan 1977), where successive observations (separated by $\Delta T$) decelerate the reversal period from $t_o$ in the "unobserved" case to an average period $\langle T \rangle$ in the observed case. In this way, an intrinsically unstable two-state system gets stabilized by its observation, so that the average reversal time $\langle T \rangle$ increases if the observation interval $\Delta T$ decreases. In the limit of continuous observation ($\Delta T \to 0$), the system becomes "frozen" in one of its possible states. Skipping the

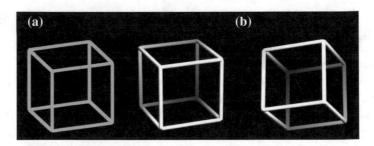

**Fig. 1   a** The Necker cube, a two-dimensional projection of a three-dimensional cube structure, as an ambiguous visual stimulus. **b** Modified cubes with depth cues removing the ambiguity of the Necker cube, so that two different, non-ambiguous stimuli are perceived

formal derivation (see Atmanspacher et al. 2004, 2008 for details), the analysis of this scenario results in the time-scale relation

$$\langle T \rangle \approx t_o^2 / \Delta T$$

between three time scales at three different orders of magnitude, which were never before studied relative to one another. These time scales are $\langle T \rangle \approx 3\,\mathrm{s}$, $t_o \approx 300\,\mathrm{ms}$, and $\Delta T \approx 30\,\mathrm{ms}$, so that the time-scale relation is satisfied.

By the time this work was done, I visited Primas at his home at Goldbach and told him about the progress we had made with this model. He listened patiently, but his response was laconic:

HP: You are not finished yet.
HA: Yes I am.
HP: How do you know all this is correct?
HA: It's been derived—why should the math fool us?
HP: But you must put it to test—experimentally!

The message was clear: although theoretical insights may provide first important steps toward progress in science, they need to be related to experiment to be ultimately convincing. This may not be obvious for a theoretician, but for someone like Primas, with his extraordinary formal *and* experimental skills, it was evident.

So, that's what we did—actually we were lucky that Jürgen Kornmeier's lab at Freiburg University had all the tools at hand that were needed to test the time-scale relation above. The trick we managed to work out was to control the time scale $t_o$ as an independent variable by presenting the Necker cube with varying off-time intervals $t_{\mathrm{off}}$. Then we could measure $\langle T \rangle$ as a function of $t_{\mathrm{off}}$ and determine $\Delta T$ from the empirical results obtained. This collaboration yielded several highly non-trivial pieces of confirmation for the time-scale relation that are shown in Fig. 2 and explained in its caption.

There is yet another important aspect of the described model that I should at least indicate, in view of Sect. 4.2 below. This aspect has to do with a temporal equivalent of entanglement correlations that may occur if system observables of temporal significance do not commute. Since this is the case for the two types of dynamics represented by $A$ and $B$, we suspected that the correlations between states at different times may violate a temporal Bell inequality first proposed by Leggett and Garg (1985).

Assuming that the perceptual system is always *uniquely* either in one or the other state, we adapted the Leggett-Garg inequality to the scenario of bistable perception and showed that it is indeed violated for particular model parameters (Atmanspacher and Filk 2010, 2013). As a consequence, it must be conjectured that the uniqueness assumption does not match the situation properly. A possible way out would be that states may be extended over time rather than being assigned to time instants with vanishing duration, resembling the idea of an extended nowness.

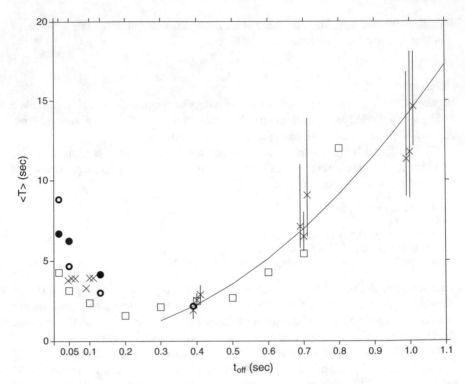

**Fig. 2** Average reversal times $\langle T \rangle$ for the bistable perception of a discontinuously presented Necker cube. Two ranges of different behavior of $\langle T \rangle$ as a function of $t_{\mathrm{off}}$ are to be distinguished: **a** $t_{\mathrm{off}} > t_0 = 300\,\mathrm{ms}$, where $t_{\mathrm{off}}$ replaces $t_o$; **b** $t_{\mathrm{off}} < t_0 = 300\,\mathrm{ms}$, where $t_o = 300\,\mathrm{ms}$ remains the relevant time scale. **a** Crosses mark results from Kornmeier and Bach (2004); for each off-time, $\langle T \rangle$ (including standard errors) is plotted for three on-times of $0.05\,\mathrm{s}$, $0.1\,\mathrm{s}$, and $0.4\,\mathrm{s}$. Squares mark results without errors indicated from Orbach et al. (1966) for an on-time of $0.3\,\mathrm{s}$. The *solid line* shows the best polynomial fit of $\langle T \rangle$ as a function of off-times $t_{\mathrm{off}}$, which is quadratic as predicted and yields $\Delta T \approx 70\,\mathrm{ms}$. **b** Empty circles are reversal times due to Kornmeier et al. (2007), crosses are results from Kornmeier and Bach (2004), and squares refer to Orbach et al. (1966). Full circles are due to simulations for assumed parameters $\Delta T = 30\,\mathrm{ms}$ and $t_0 = 300\,\mathrm{ms}$ as in Atmanspacher et al. (2008)

However, a violation of a temporal Bell inequality is difficult to realize experimentally (and, in fact, hasn't been realized so far): any measurement at one time potentially induces local correlations with any later measurement. Therefore a violation remains inconclusive if such "invasive" measurements cannot be excluded—or correlations due to invasivity cannot be distinguished from genuine entanglement correlations (cf. Dzhafarov and Kujala 2013).

# 4 Dual-Aspect Monism

## 4.1 The Pauli–Jung Conjecture

One of the long-standing interests of Hans Primas[8] was the interaction between the physicist Wolfgang Pauli and the psychologist Carl Gustav Jung between 1932 and 1958. When Pauli's correspondence with Jung and many others was published in eight successive volumes between 1979 and 2005, Primas apparently read everything that touched the *psychophysical problem*, the term that Pauli and Jung used for the problem of the relationship between the mental and the material (Fig. 3).

As this was my first point of contact with Primas back in 1991, it is not surprising that the Pauli-Jung dialog, as we called it early on, seriously occupied both of us and gave rise to conferences and workshops that we jointly organized. Over the years, we were able to reconstruct a consistent picture of their ideas, which they never published in a coherent framework, and discovered that it matches the broad class of dual-aspect monist approaches to the psychophysical problem (Atmanspacher et al. 1995; Atmanspacher and Primas 1996, 2006, 2009).

The gist of dual-aspect monism is the idea to combine an epistemic dualism of the mental and the material with an ontic monism of an underlying, psychophysically

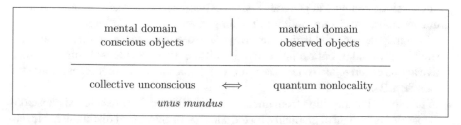

**Fig. 3** In dual-aspect monism according to Pauli and Jung, the mental and the material are manifestations of an underlying, psychophysically neutral, holistic reality, called *unus mundus*, whose symmetry must be broken to yield dual, complementary aspects. From the mental the neutral reality is approached via Jung's collective unconscious, from the material it is approached via quantum nonlocality

---

[8]In his meticulous biographical notes Primas indicates, almost indiscernably hidden among references to oodles of books on science and engineering, an awakening interest for consciousness and the unconscious in November 1944, as a 16-year-old. A year later he became fascinated with Jung's *Psychology and Alchemie* in 1945 (translation by HA): "The impact of this book—which I read only partially and diagonally at the time—was peculiar and lasting... It was striking that Jung's thoughts, somewhat odd relative to my materialistically shaped mindset, convinced me immediately, as if I had long foreboded them." In the following years Primas continued his studies of Jung's works until he began visiting lectures at ETH in 1949.

neutral domain. This general idea has variants though. Five specific features of the conjecture proposed by Pauli and Jung are the following[9]:

- Applying the ontic-epistemic distinction to the Pauli-Jung framework of thinking reflects that the mental and the material are basically regarded as modes of knowledge acquisition about something ontic which is itself not epistemically— i.e. empirically—accessible (compare Sect. 2.2). This is particularly relevant for Jung's notion of archetypal patterns: while archetypes themselves are conceived as structural ordering principles within the psychophysically neutral ontic domain, archetypal patterns and images appear as their mental manifestations, subject to concrete experience.
- At the fundament of the ontic level, reality is undivided, distinction-free (cf. the notion of the "unidived universe" by Bohm and Hiley; see Hiley in this volume). This illustrates why, in the limit of such an *unus mundus*, epistemic access to the ontic is impossible: if there are no distinctions, there are no categories to be distinguished. The move from the *unus mundus* via Jung's collective unconscious to refined mental categories, or via a nonlocal physical reality to physical objects, is *decompositional*. This is different from Russell's neutral monism or Chalmers' naturalistic dualism, where mental and material objects arise due to *compositions* of psychophysically neutral elements.
- The absolute impossibility of epistemic access (a neo-Kantian feature in late Jungian thinking) strictly applies to the undivided *unus mundus* only, not to all unconscious contents in general. Every distinction that is made, even within the unconscious, creates the option of forming categories (e.g. different archetypes), which may be accessed if there are ways to experience them. In this sense, each such level would be ontic relative to more differentiated levels, and epistemic relative to less differentiated ones. The ontic-epistemic distinction becomes relativized (see Sect. 2.4).
- In this spirit, the transition from unconscious activity to fully developed conscious categories is thought of gradually, by the successive creation of distinguishable features, which Pauli (1954) speculated to be analogous to physical measurement (see Sect. 2.3). The process by which unconscious contexts are transformed into consciousness is *active* insofar as it includes a reaction back onto the unconscious,[10] just as measurement in physics changes the measured physical state. This idea is decisive for the application of non-commutative structures to psychology and cognitive science, outlined in Sect. 3.
- The dual aspects in the Pauli-Jung conjecture are understood as *complementary* (see Sect. 3.1). This means that the corresponding epistemic perspectives of the mental and the material exclude one another in the sense of a logical exclusive or

---

[9]The notion of the "Pauli-Jung conjecture" emerged in the early 2010s, when it became clear that dual-aspect monism à la Pauli and Jung entails a number of ramifications that have empirically testable consequences (see Atmanspacher and Fach 2013).

[10]Otherwise, the whole purpose of psychotherapy or -analysis as a method to change unconscious roots of conscious symptoms would be pointless. The active backreaction also casts the popular notion of consciousness as a mere filter into doubt.

("either–or"). At the level of the psychophysically neutral, the logical negation of the exclusive or applies ("neither–nor"), because this level does not contain the distinction between the mental and the material. This must not be confused with the logical inclusive or ("both–and").

In a broader picture, Pauli's and Jung's ideas were outstanding in yet another sense: in an intellectual climate of a clear move toward the rejection of ontology and metaphysics in the 20th century, they postulated exactly the opposite: that metaphysical assumptions are mandatory and even useful if one wants to address questions of basic relevance.[11] After the logical empiricism of the Vienna circle, after Bohr's epistemic standpoint in quantum physics, and after the linguistic turn initiated by Wittgenstein, Pauli und Jung suggested that we need a completely new idea of reality, which exceeds our theories about nature in particular and language in general. Jung's emphasis on the "reality of the symbol", very much welcomed by Pauli, may be an important issue in this respect that should be explored further.

## 4.2  Mental and Physical Time

As much as Hans Primas was interested in dual-aspect monism as an approach toward the psychophysical problem, he was well aware that speculative metaphysical ideas alone will not have the power of transforming a world view. Therefore, he spent much time in his last two decades to explore novel avenues along which one might hope for more concrete insights. His strategic move was to acuminate the psychophysical problem as a whole down to a facet of it that may be restrictive enough to give us hints for where scientific progress toward a better understanding might be possible.

The fact that he identified is itself one of the great problems throughout the history of ideas: the problem of how mental and physical aspects of time are related to one another. In two publications, Primas (2003, 2009) sketched a way in which *temporal entanglement* might be a key to unlock several riddles behind mental and physical time. There is a lot of philosophical literature about them, much of which bases their distinction on the notion of tense.

In physics, the fundamental laws of motion (or their solutions, respectively) are time-translation invariant, time-reversal invariant, and time-scale invariant. These invariances, also called symmetries, mean that physical time at the fundamental level has no privileged instant (present), no preferred direction (past or future), and no intrinsic scale (time unit). The only relation between two instants in physical *tenseless time* is that their values on a time axis are greater or smaller than the other.

Mental time, on the other hand, features the tenses—past, present, and future— as key notions. So, at least time-translation symmetry and time-reversal symmetry are broken by mental *tensed time*. Moreover, the phenomenological experience of

---

[11]In this context, Primas liked to cite Carl Friedrich von Weizsäcker (in personal conversation): "Every scientist works with metaphysical assumptions, and those who deny this most usually work with the poorest ones."

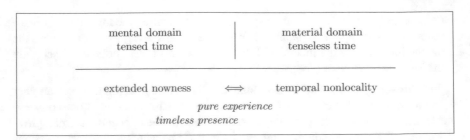

**Fig. 4** Restriction of dual-aspect monism to mental and physical time as suggested by Primas. From the mental the neutral reality is approached via the experience of "extended nowness", from the material it is approached via the concept of "temporal nonlocality"

time suggests that the present is not an extensionless instant between past and future but has internal duration, an extended nowness as it were. Philosophers have coined concepts such as the "specious present" (James) or "actual occasion" (Whitehead) to take this into account (Fig. 4).

It is plausible to consider the experience of an extended nowness as the most elementary kind of phenomenal content (quale) of a mental state, without which no other qualia experience can possibly be made. In this sense, nowness is the basis of all experience. James' notion of "pure experience", his way of addressing the psychophysically neutral, resonates with this fundamental mode of the experience of presence in the present.[12]

The physicist's way to enter the domain of psychophysically neutral nowness proceeds via temporal nonlocality as referred to at the end of Sect. 3.2. The idea here is that pieces of nowness exist between successive elementary events, say $e_1$ and $e_2$, so that nothing in the interval between them could be used to define any time ordering or, for the same reason, causal relations within this interval. For the level of the *unus mundus* this implies a completely timeless presence, because there are no distinguishable events at all.

George Sudarshan, who together with Gustafson and Misra pioneered the quantum Zeno effect indicated in Sect. 3.2, once posed the question of whether we can "perceive a quantum system directly", and speculated about a mode of awareness in which (Sudarshan 1983)

> sensations, feelings, and insights are not neatly categorized into chains of thoughts, nor is there a step-by-step development of a logical-legal argument-to-conclusion. Instead, patterns appear, interweave, coexist; and sequencing is made inoperative. Conclusion, premises, feelings, and insights coexist in a manner defying temporal order.

The visionary outlook of Hans Primas relates all these ideas to the framework of thinking developed by Pauli and Jung. It does so in his typical manner, heavily relying on mathematical concepts couched in algebraic and group theoretical language and

---

[12]"Pure experience" is an ambiguous term, however, since it triggers a mental understanding, similar to Jung's term "archetypal image" if it were used for the psychophysically neutral.

based on his expert knowledge of engineering mathematics. In his first explicit text about time entanglement he states (Primas 2003):

> Our point of departure is the hypothesis that there is a timeless holistic reality which can be described in the non-Boolean logical structure of modern quantum theory. Neither time, nor mind, nor matter and energy, are taken to be a priori concepts. Rather it is assumed that these concepts emerge by a contextual breaking of the holistic symmetry of the unus mundus.

In his final publication (Primas 2009), which Primas saw as an essential refinement of the 2003 paper, he introduces the affine Weyl-Heisenberg group to implement the three time symmetries and their breakdown. The resulting subgroups lead into the domains of tensed mental time and tenseless physical time (often called "A-time" and "B-time" in the philosophical literature):

> The traditional difficulties with the concepts "A-time" and "B-time" arise because they cannot be captured within a single Boolean description. But they can be conceived in terms of a non-Boolean description generated by the affine Weyl-Heisenberg symmetry group. Epistemically accessible partial descriptions can then be generated by an epistemic breaking of the full temporal symmetry. The two affine subgroups of the affine Weyl-Heisenberg group are complementary in a mathematically well-defined sense and allow a precise description of A-time and B-time, respectively. It follows that both A-time and B-time are necessary but none of them has a privileged status, none of them can replace the other.

In the years before he died in 2014, Primas continued to revise and expand his views and ideas on time, mind and matter in a 600-pages manuscript that he left in a fairly complete but unedited state. This manuscript will soon be published under the title *Knowledge and Time*. One could not think of a better testimony for a scholar who spent his scientific life on an avenue so unusual, and at the same time so coherent, as the path of Hans Primas: from engineering and chemistry to the foundations of physics and to the metaphysics of consciousness—stimulation and inspiration for everyone who has the thirst for insight and the intellectual freedom to follow.

# References

Aerts, D., and Aerts S. (1995): Applications of quantum statistics in psychological studies of decision processes. *Foundations of Science* **1**, 85–97.

Aerts, S., Kitto, K., and Sitbon L. (2011): Similarity metrics within a point of view. In *Quantum Interaction QI 15*, ed. by D. Song *et al.*, Springer, Berlin, pp. 13–24.

Allahverdyan, A.E., Balian, R., and Niewenhuizen, T.M. (2013): Understanding quantum measurement from the solution of dynamical models. *Physics Reports* **525**, 1–166.

Amann, A., and Atmanspacher, H. (1999): C*- and W*-algebras of observables, their interpretations, and the problem of measurement. In *On Quanta, Mind and Matter*, ed. by H. Atmanspacher, A. Amann, U. Müller-Herold, Kluwer, Dordrecht, pp. 57–79.

Amann, A. and Primas, H. (1997): What is the referent of a non-pure quantum state? In *Potentiality, Entanglement, and Passion-at-a-Distance*, ed. by R.S. Cohen, M. Horne, and J. Stachel, Kluwer, Dordrecht, pp. 9–29.

Antoniou, I., Gialampoukidis, I., and Ioannidis, E. (2016): Age and time operator of evolutionary processes. In *Quantum Interaction QI 15*, ed. by H. Atanspacher *et al.*, Springer, Berlin, pp. 31–75.

Atmanspacher, H. (2015): Quantum aproaches to consciousness. *Stanford Encyclopedia of Philosophy*, ed. by E.N. Zalta. Accessible at http://plato.stanford.edu/entries/qt-consciousness/.

Atmanspacher, H., Bach, M., Filk, T., Kornmeier, J., & Römer, H. (2008): Cognitive time scales in a Necker-Zeno model for bistable perception. *Open Cybernetics and Systemics Journal 2*, 234–251.

Atmanspacher, H., Bezzola Lambert, L., Folkers, G., and Schubiger, P.A. (2014): Relevance relations for the concept of reproducibility. *Journal of the Royal Society Interface* **11**(94), 20131030.

Atmanspacher, H., and Fach, W. (2013): A structural-phenomenological typology of mind-matter correlations. *Journal of Analytical Psychology* **58**, 219–244.

Atmanspacher, H., Filk T., and Römer, H. (2004): Quantum Zeno features of bistable perception. *Biological Cybernetics 90*, 33–40.

Atmanspacher, H., and Filk T. (2006): Complexity and non-commutativity of learning operations on graphs. *BioSystems* **85**, 84–93.

Atmanspacher, H., and Filk T. (2010): A proposed test of temporal nonlocality in bistable perception. *Journal of Mathematical Psychology* **54**, 314–321.

Atmanspacher, H., and Filk T. (2013): The Necker-Zeno model for bistable perception. *Topics in Cognitive Science 5*, 800–817.

Atmanspacher, H., and beim Graben, P. (2016): Incompatible descriptions of systems across scales of granularity. In *Quantum Interaction QI 15*, ed. by H. Atanspacher *et al.*, Springer, Berlin, pp. 113–125.

Atmanspacher, H., and Kronz, F. (1999): Relative onticity. In *On Quanta, Mind and Matter*, ed. by H. Atmanspacher, A. Amann, U. Müller-Herold, Kluwer, Dordrecht, pp. 273–294.

Atmanspacher, H., Primas, H., and Wertenschlag, E., eds. (1995): *Der Pauli-Jung-Dialog und seine Bedeutung für die moderne Wissenschaft*, Springer, Berlin.

Atmanspacher, H., and Primas, H. (1996): The hidden side of Wolfgang Pauli. *Journal of Consciousness Studies* **3**(2), 112–126. Reprinted by several other journals.

Atmanspacher H. and Primas H. (2003): Epistemic and ontic quantum realities. In *Time, Quantum and Information*, ed. by L. Castell and O. Ischebeck, Springer, Berlin, pp. 301-321.

Atmanspacher, H., and Primas, H. (2006): Pauli's ideas on mind and matter in the context of contemporary science. *Journal of Consciousness Studies* **13**(3), 5–50.

Atmanspacher, H., and Primas, H., eds. (2009): *Recasting Reality. Wolfgang Pauli's Philosophical Ideas and Contemporary Science*, Springer, Berlin.

Atmanspacher, H., and Römer H. (2012): Order effects in sequential measurememts of non-commuting psychological observables. *Journal of Mathematical Psychology* **56**, 274–280.

Bishop, R.C., and Atmanspacher, H. (2006): Contextual emergence in the description of properties. *Foundations of Physics* **36**, 1753–1777.

Bohr, N. (1949): Discussion with Einstein on epistemological problems in atomic physics. In *Albert Einstein: Philosopher–Scientist*, ed. by P.A. Schilpp, Library of Living Philosophers, Evanston, pp. 199–241.

Bruza, P.D., Kitto, K., Ramm, B.R., and Sitbon L. (2015): A probabilistic framework for analysing the compositionality of conceptual combinations. *Journal of Mathematical Psychology* **67**, 26–38.

Busemeyer, J.R., and Bruza, P.D. (2013): *Quantum Models of Cognition and Decision*, Cambridge University Press, Cambridge.

Dzhafarov, E.N., and Kujala, J.V. (2013): All-possible-couplings approach to measuring probabilistic context. *PLoS One* **8**(5): e61712.

Einstein, A. (1949a): Autobiographical notes. In *Albert Einstein: Philosopher–Scientist*, ed. by P.A. Schilpp, Library of Living Philosophers, Evanston, pp. 1–95.

Einstein, A. (1949b): Reply to criticism. In *Albert Einstein: Philosopher–Scientist*, ed. by P.A. Schilpp, Library of Living Philosophers, Evanston, pp. 665–688.

Gabora, L., and Aerts D. (2002): Contextualizing concepts using a mathematical generalization of the quantum formalism. *Journal of Experimental and Theoretical Artificial Intelligence* **14**, 327–358.

Garfinkel, A. (1981): *Forms of Explanation*, Yale University Press, New Haven.

beim Graben, P., and Atmanspacher, H. (2006): Complementarity in classical dynamical systems. *Foundations of Physics* **36**, 291–306.

Gustafson, K., and Misra, B. (1976): Canonical commutation relations of quantum mechanics and stochastic regularity. *Letters in Mathematical Physics* **1**, 275–280.

Hepp, K. (1972): Quantum theory of measurement and macroscopic observables. *Helvetica Physica Acta* **45**, 237–248.

Kornmeier, J., and Bach, M. (2004): Early neural activity in Necker cube reversal: Evidence for low-level processing of a gestalt phenomenon. *Psychophysiology* **41**, 1–8.

Kornmeier, J., Ehm, W., Bigalke, H., and Bach, M. (2007). Discontinuous presentation of ambiguous figures: How interstimulus-interval durations affect reversal dynamics and ERPs. *Psychophysiology* **44**, 552–560.

Leggett, A.J., and Garg, A. (1985): Quantum mechanics versus macroscopic realism: Is the flux there when nobody looks? *Physical Review Letters* **54**, 857–860.

Lockhart, C.M., and Misra, B. (1986): Irreversibility and measurement in quantum mechanics. *Physica A* **136**, 47–76.

Misra, B., and Sudarshan, E.C.G. (1977): The Zeno's paradox in quantum theory. *Journal of Mathematical Physics* **18**, 756–763.

Orbach, J., Zucker, E., and Olson, R. (1966). Reversibility of the Necker cube: VII. Reversal rate as a function of figure-on and figure-off durations. *Perceptual and Motor Skills* **22**, 615–618.

Pauli, W. (1954): Letter to Jung. In *On the Nature of the Psyche* by Jung (1969), Collected Works Vol. 8, pp. 159–236 (footnote 130), Princeton University Press, Princeton.

Petersen, A. (1963): The philosophy of Niels Bohr. *Bulletin of the Atomic Scientist* **19**(7), 8–14.

Popescu, S., and Rohrlich, D. (1994): Nonlocality as an axiom. *Foundations of Physics* **24**, 379–385.

Pothos, E.M., and Busemeyer, J.R. (2009): A quantum probability model explanation for violations of rational decision theory. *Proceedings of the Royal Society B* **276**, 2171–2178.

Prigogine, I. (1980): *From Being to Becoming*, Freeman, San Francisco.

Primas, H. (1977): Theory reduction and non-Boolean theories. *Journal of Mathematical Biology* **4**, 281–301.

Primas H. (1990): Mathematical and philosophical questions in the theory of open and macroscopic quantum systems. In *Sixty-Two Years of Uncertainty*, ed. by A.I. Miller, Plenum, New York, pp. 233–257.

Primas H. (1991): Necessary and sufficient conditions for an individual description of the measurement process. In *Symposium on the Foundations of Modern Physics 1990*, ed. by P. Lahti and P. Mittelstaedt, World Scientific, Singapore, pp. 332–346.

Primas H. (1993): The Cartesian cut, the Heisenberg cut, and disentangled observers. In *Symposia on the Foundations of Modern Physics 1992*, ed. by K.V. Laurikainen and C. Montonen, World Scientific, Singapore, pp. 245–269.

Primas H. (1994a): Endo- and exo-theories of matter. In *Inside Versus Outside*, ed. by H. Atmanspacher and G.J. Dalenoort, Springer, Berlin, pp. 163–193.

Primas H. (1994b): Hierarchic quantum descriptions and their associated ontologies. *Symposium on the Foundations of Quantum Mechanics 1994*, ed. by K.V. Laurikainen, C. Montonen and K. Sunnarborg, Editions Frontières, Gif-sur-Yvette, pp. 210–220.

Primas (1997): Individual description of dynamical state reductions in quantum mechanics. Unpublished manuscript.

Primas, H. (1998): Emergence in exact natural sciences. *Acta Polytechnica Scandinavica* **Ma 91**, 83–98.

Primas (2003): Time-entanglement between mind and matter. *Mind and Matter* **1**, 81–119.

Primas, H. (2007): Non-Boolean descriptions for mind-matter problems. *Mind and Matter* **5**, 7–44.

Primas, H. (2009): Complementarity of mind and matter. In *Recasting Reality*, ed. by H. Atmanspacher and H. Primas, Springer, Berlin, pp. 171–209.

Putnam, H. (1981): *Reason, Truth and History*. Cambridge University Press, Cambridge.

Quine, W.V. (1969): Ontological relativity. In *Ontological Relativity and Other Essays*, Columbia University Press, New York, pp. 26–68.

Scheibe, E. (1964): *Die kontingenten Aussagen in der Physik*, Athenäum, Frankfurt.

Scheibe, E. (1973): *The Logical Analysis of Quantum Mechanics*, Pergamon, Oxford.

Shimony, A. (1993): The transient now. In *Search for a Naturalistiv Worldview, Vol. 2*, Cambridge University Press, Cambridge, pp. 271–288.

Sudarshan, E.C.G. (1983) Perception of quantum systems. In *Old and New Questions in Physics, Cosmology, Philosophy, and Theoretical Biology*, ed. by A. van der Merwe, Plenum, New York, pp. 457–467.

van Fraassen, B. (1980): *The Scientific Image*. Clarendon, Oxford.

Wang, Z., Solloway, T., Shiffrin, R.M., and Busemeyer, J.R. (2014): Context effects produced by question orders reveal quantum nature of human judgments. *Proceedings of the National Academy of Sciences of the USA* **111**, 9431–9436.

Wendt, A. (2015): *Quantum Mind and Social Science*, Cambridge University Press, Cambridge.

# Publications by Hans Primas

Osimitz F. and Primas H. (1950): Tüpfelreaktionen. *Schweizerische Laboranten-Zeitung* **7**, 2–7.

Primas H., Lasman H., and Osimitz F. (1950): Moderne Vorschriften zur qualitativen Kationenanalyse. *Schweizerische Laboranten-Zeitung* **7**, 98–114.

Primas H. and Günthard Hs.H. (1953): Die Infrarotspektren von Kettenmolekülen der Formel R'CO(CH"CH")$_n$COR". I. Rocking- und Twisting-Grundtöne. *Helvetica Chimica Acta* **36**, 1659–1670.

Primas H. and Günthard Hs.H. (1953): Die Infrarotspektren von Kettenmolekülen der Formel R'CO(CH"CH")$_n$COR". II. Die Normalschwingungen des Symmetrietypus $B_u$. *Helvetica Chimica Acta* **36**, 1791–1803.

Primas H. and Günthard Hs.H. (1954): Spectres infrarouges de derivés carbonyliques du type R'CO(CH"CH")$_n$COR" contenant plus de dix groupes méthyléniques. *Journal de Physique et le Radium* **15**, 209–211.

Primas H. and Günthard Hs.H. (1954): Theorie der Form von Absorptionsbanden suspendierter Substanzen und deren Anwendung auf die Nujolmethode in der Infrarotspektroskopie. *Helvetica Chimica Acta* **37**, 360–374.

Primas H. and Günthard Hs.H. (1955): Theorie der Intensitäten der Schwingungsspektren von Kettenmolekeln. I. Allgemeine Theorie der Berechnung von Intensitäten der Infrarotspektren von grossen Molekeln. *Helvetica Chimica Acta* **38**, 1254–1262.

Primas H. and Günthard Hs.H. (1956): Theorie der Intensitäten der Schwingungsspektren von Kettenmolekeln. II. Zur Berechnung der Intensitäten der Infrarotspektren von freien Kettenmolekeln der Symmetrie C$_{2h}$. *Helvetica Chimica Acta* **39**, 1182–1192.

Günthard Hs.H. and Primas H. (1956): Zusammenhang von Graphentheorie und MO-Theorie von Molekeln mit Systemen konjugierter Bindungen. *Helvetica Chimica Acta* **39**, 1645–1653.

Primas H. (1957): Ein Kernresonanzspektrograph mit hoher Auflösung. I. Theorie der Liniendeformation in der hochauflösenden Kernresonanzspektroskopie. *Helvetica Physica Acta* **30**, 297–314.

© Springer International Publishing Switzerland 2016
H. Atmanspacher and U. Müller-Herold (eds.), *From Chemistry to Consciousness*, DOI 10.1007/978-3-319-43573-2

Primas H. and Günthard Hs.H. (1957): Ein Kernresonanzspektrograph mit hoher Auflösung. II. Beschreibung der Apparatur. *Helvetica Physica Acta* **30**, 315–330.

Primas H. and Günthard Hs.H. (1957): Herstellung sehr homogener axialsymmetrischer Magnetfelder. *Helvetica Physica Acta* **30**, 331–346.

Primas H. and Günthard Hs.H. (1957): Field stabilizer for high resolution nuclear magnetic resonance. *Review of Scientific Instruments* **28**, 510–514.

Primas H. and Günthard Hs.H. (1957): Hochauflösender Kernresonanzspektrograph. *Chimia* **11**, 130–132.

Primas H., Frei K., and Günthard Hs.H. (1958): Protonenresonanzspektren einfacher cyclischer Aether und Ketone I. *Helvetica Chimica Acta* **41**, 35–38.

Primas H. (1958): Ein Modulationsverfahren für die Kernresonanzspektroskopie hoher Auflösung. *Helvetica Physica Acta* **31**, 17–24.

Primas H. and Günthard Hs.H. (1958): Eine Methode zur direkten Berechnung des Spektrums der von quantenmechanischen Systemen absorbierten bzw. emittierten elektromagnetischen Strahlung. *Helvetica Physica Acta* **31**, 413–434.

Primas H. (1959): A new method for analyzing spectra in high resolution NMR spectroscopy. In *Proceedings of the Conference of Molecular Spectroscopy*, ed. by R. Thornton and H.W. Thompson, Pergamon, London, pp. 19–25.

Primas H. (1959): Anwendungen der magnetischen Kernresonanz in der Chemie. *Chimia* **13**, 15–23.

Primas H., Arndt R., and Ernst R. (1959): Die Konstruktion von Kernresonanz-Spektrographen hoher Auflösung Ia. *Zeitschrift für Instrumentenkunde* **67**, 293–300.

Primas H., Arndt R., and Ernst R. (1960): Die Konstruktion von Kernresonanz-Spektrographen hoher Auflösung Ib. *Zeitschrift für Instrumentenkunde* **68**, 8–13.

Primas H., Arndt R., and Ernst R. (1960): Die Konstruktion von Kernresonanz-Spektrographen hoher Auflösung. II. Die Konstruktion des Hochfrequenzteiles von Kernresonanz-Spektrographen hoher Auflösung. *Zeitschrift für Instrumentenkunde* **68**, 21–29.

Primas H., Arndt R., and Ernst R. (1960): Die Konstruktion von Kernresonanz-Spektrographen hoher Auflösung. III. Einige aktuelle Probleme der Kernresonanz-Instrumentierung. *Zeitschrift für Instrumentenkunde* **68**, 55–62.

Primas H. (1961): Über quantenmechanische Systeme mit einem stochastischen Hamiltonoperator. *Helvetica Physica Acta* **34**, 36–57.

Primas H. (1961): Eine verallgemeinerte Störungstheorie für quantenmechanische Mehrteilchenprobleme. *Helvetica Physica Acta* **34**, 331–351.

Primas H. (1962): 35 Jahre Quantenchemie. *Chimia* **16**, 281–289.

Primas H., Arndt R., and Ernst R. (1962): Group contributions to the chemical shift in proton magnetic resonance of organic compounds. In *Advances in Molecular Spectroscopy*, ed. by A. Mangini, Pergamon, Oxford, pp. 1246–1252.

Ernst R. and Primas H. (1962): High resolution NMR-instrumentation: Recent advances and prospects. *Discussions of the Faraday Society* **34**, 43–51.

Ernst R. and Primas H. (1963): Nuclear magnetic resonance with stochastic high-frequency fields. *Helvetica Physica Acta* **36**, 583–600.

Primas H. (1963): Generalized perturbation theory in operator form. *Reviews of Modern Physics* **35**, 710–712.

Banwell C.N. and Primas H. (1963): On the analysis of high-resolution nuclear magnetic resonance spectra. I. Methods of calculating NMR spectra. *Molecular Physics* **6**, 225–256.

Ernst R. and Primas H. (1963): Gegenwärtiger Stand und Entwicklungstendenzen in der Instrumentierung hochauflösender Kernresonanz-Spektrometer. *Berichte der Bunsengesellschaft für physikalische Chemie* **67**, 261–267.

Primas H. (1964): Was sind Elektronen? *Helvetica Chimica Acta* **47**, 1840–1851.

Huber A. and Primas H. (1965): On the design of wide range electromagnets of high homogeneity. *Nuclear Instruments and Methods* **33**, 125–130.

Primas H. (1965): Separability in many-electron systems. *Modern Quantum Chemistry, Part 2*, ed. by O. Sinanoglu, Academic, New York, pp. 45–74.

Primas H. (1965): Was sind Elektronen? *Chimia* **19**, 399.

Primas H. und Riess J. (1966): Linear diamagnetic and paramagnetic response. In *Quantum Theory of Atoms, Molecules, and the Solid State. A Tribute to John C. Slater*, ed. by P.O. Löwdin, Academic, New York, pp. 319–333.

Günthard Hs. H. and Primas H. (1967): Prof. Dr. A. Stieger, 80jährig. *Titania Winterthur AHC*, 5–6.

Primas H. (1967): A density functional representation of quantum chemistry. I. Motivation and general formalism. *International Journal of Quantum Chemistry* **1**, 493–519.

Primas H. (1968): Zur Theorie grosser Molekeln. I. Revision der Grundlagen der Quantenchemie. *Helvetica Chimica Acta* **51**, 1037–1051.

Riess J. und Primas H. (1968): A variational principle for the phase of the wave function of molecular systems. *Chemical Physics Letters* **1**, 545–548.

Primas H. (1968): Lars Onsager – ein Meister der theoretischen Chemie. *Neue Züricher Zeitung*, Nr. 700, p. 5.

Primas H. (1973): Chemische Bindung. Ausgearbeitet von A. Wokaun, Verlag der Fachvereine, Zürich.

Primas H. and Schleicher M. (1975): A density functional representation of quantum chemistry. II. Local quantum theories of molecular matter in terms of the charge density operator do not work. *International Journal of Quantum Chemistry* **9**, 855–870.

Schleicher M. and Primas H. (1975): A density functional representation of quantum chemistry. III. Rigorous realization of the program in lattice space. *International Journal of Quantum Chemistry* **9**, 871–886.

Primas H. (1975): Pattern recognition in molecular quantum mechanics. I. Background dependence of molecular states. *Theoretica Chimica Acta* **39**, 127–148.

Primas H. (1976): Gibt es eine theoretische Chemie? In *Studienführer Chemie*, hrsg. von der Vereinigung der Assistenten an den chemischen Laboratorien der ETH Zürich, pp. 85–87.

Primas H. (1977): Theory reduction and non-Boolean theories. *Journal of Mathematical Biology* **4**, 281–301.

Primas H. and Müller-Herold U. (1978): Quantum mechanical system theory: A unifying framework for observations and stochastic processes in quantum mechanics. *Advances in Chemical Physics* **38**, 1–107.

Saraswati D.K. and Primas H. (1978): A system theoretic representation of mechanical systems. I. Decomposition of a mechanical system into a hierarchy of orthogonal stationary linear dynamical systems. *Journal of Mathematical Physics* **19**, 2646–2654.

Saraswati D.K. and Primas H. (1978): A system theoretic representation of mechanical systems. II. Stochastic interpretation. *Journal of Mathematical Physics* **19**, 2655–2658.

Primas H. (1978): Kinematical symmetries in molecular quantum mechanics. In *Lecture Notes in Physics* **79**, Springer, Berlin, pp. 72–91.

Primas H. (1978): *Elemente der Gruppentheorie*, Verlag der Fachvereine, Zürich.

Primas H. (1979): Chemie, Reduktionismus und Quantenlogik. *Match* **7**, 217.

Primas H. and Gans W. (1979): Quantenmechanik, Biologie und Theoriereduktion. In *Materie–Leben–Geist. Zum Problem der Reduktion der Wissenschaften*, hrsg. von B. Kanitscheider, Duncker & Humblot, Berlin, pp. 15–42.

Primas H. (1980): Foundations of theoretical chemistry. In *Quantum Dynamics of Molecules. The New Experimental Challenge to Theorists*, ed. by R.G. Woolley, Plenum, New York, pp. 39–113.

Primas H. (1981): *Chemistry, Quantum Mechanics and Reductionism*, Springer, Berlin.

Primas H. (1982): Chemistry and complementarity. *Chimia* **36**, 293–300.

Raggio G.A. and Primas H. (1982): Remarks on "On completely positive maps in generalized quantum dynamics". *Foundations of Physics* **12**, 433–435.

Primas H. (1983): Quantum mechanics and chemistry. In *Les Fondements de la Mécanique Quantique*, ed. by C. Gruber, C. Piron, T.M. Tâm, A. V. C. P., Lausanne, pp. 255–270.

Primas H. (1984): Verschränkte Systeme und Komplementarität. In *Moderne Naturphilosophie*, hrsg. von B. Kanitscheider, Königshausen & Neumann, Würzburg, pp. 243–260.

Primas H. (1984): Concepts of quantum theory. Review of *Quantum Theory and Measurement* by J.A. Wheeler and W.H. Zurek (Princeton University Press), *Nature* **308**, 782–783.

Primas H. and Müller-Herold U. (1984): *Elementare Quantenchemie*, Teubner, Stuttgart.

Primas H. (1984): Was ist die Teilchenforschung wert? Beitrag zu einer Podiumsdiskussion. In *Bild der Wissenschaft* **21**(9), 126–141.

Primas H. (1985): Kann Chemie auf Physik reduziert werden? *Neue Zürcher Zeitung*, Nr. 42, 67–68.

Primas H. (1985): Kann Chemie auf Physik reduziert werden? Erster Teil: Das molekulare Programm, zweiter Teil: die Chemie der Makrowelt. *Chemie in unserer Zeit* **19**, 109–119, 160–166.

Primas H. (1986): Die Quantenmechanik als umfassende Theorie der Realität. In *Europäisches Forum Alpbach*, hrsg. von O. Molden, Österreichisches College, Wien, pp. 140–155.

Primas H. (1987): Objekte in der Quantenmechanik. In *Grazer Gespräche 1986: Ganzheitsphysik*, hrsg. von M. Heindler und F. Moser, Technische Universität, Graz, pp. 163–201.

Primas H. (1987): Contextual quantum objects and their ontic interpretation. In *Symposium on the Foundations of Modern Physics 1987*, ed. by P. Lahti and P. Mittelstaedt, World Scientific, Singapore, pp. 251–275.

Primas H. (1988): Zum Theorienpluralismus in den Naturwissenschaften. In *Wozu Wissenschaftsphilosohie?* Hrsg. von P. Hoyningen-Huene und G. Hirsch, de Gruyter, Berlin, pp. 172–178.

Primas H. (1988): The essentials of the Copenhagen interpretation. Contribution to a panel discussion. In *Symposium on the Foundations of Modern Physics 1987*. Report Series Turku-FTL-L45, Department of Physical Sciences, University of Turku, pp. 17–21 and p. 83.

Primas H. (1988): Visszavezethetö-e a kémia a fizikára? *Mérlwg* **88/3**, 247–266.

Primas H. (1988): Can we reduce chemistry to physics? In *Centripetal Forces in the Sciences, Vol. II*, ed. by G. Radnitzky, Paragon House, New York, pp. 119–133.

Primas H. (1988): Rebuttal to the comments by Marcelo Alonso. In *Centripetal Forces in the Sciences, Vol. II*, ed. by G. Radnitzky, Paragon House, New York, pp. 137–143.

Primas H. (1989): Great expectations. Review of *Beyond the Atom. The Philosophical Thought of Wolfgang Pauli* by K.V. Laurikainen (Springer, Berlin), *Nature* **338**, pp. 305–306.

Primas H. (1990): Biologie ist mehr als Molekularbiologie. In *Die Frage nach dem Leben*, hrsg. von E.P. Fischer und K. Mainzer, Piper, München, pp. 63–92. Partly reprinted in *Einzelmaterial 6, Sek. II, 6*, Biologie, November 1995.

Primas H. (1990): Zur Quantenmechanik makroskopischer Systeme. In *Wieviele Leben hat Schrödingers Katze?* Hrsg. von J. Audretsch und K. Mainzer, Wissenschaftsverlag, Mannheim, pp. 207–243.

Primas H. (1990): Realistic interpretation of the quantum theory for individual objects. *La Nuova Critica* **13–14**, pp. 41–72.

Primas H. (1990): Mathematical and philosophical questions in the theory of open and macroscopic quantum systems. In *Sixty-Two Years of Uncertainty*, ed. by A.I. Miller, Plenum, New York, pp. 233–257.

Primas H. (1990): Induced nonlinear time evolution of open quantum objects. In *Sixty-Two Years of Uncertainty*, ed. by A.I. Miller, Plenum, New York, pp. 259–280.

Primas H. (1990): The measurement process in the individual interpretation of quantum mechanics. In *Quantum Theory without Reduction*, ed. by M. Cini and J.M. Lévy-Leblond, Adam Hilger, Bristol, pp. 49–68.

Primas H. (1990): Beyond Baconian quantum physics. In *Kohti uutta todellisuuskäsitystä*, Yliopistopaino, Helsinki, pp. 100–112.

Primas H. (1991): Besprechung von *The Philosophy of Quantum Mechanics* von R. Healey (Cambridge University Press). *Physikalische Blätter* **47**, Nr. 1, 68.

Primas H. (1991): Necessary and sufficient conditions for an individual description of the measurement process. In *Symposium on the Foundations of Modern*

*Physics 1990*, ed. by P. Lahti and P. Mittelstaedt, World Scientific, Singapore, pp. 332–346.

Primas H. (1991): Remarks on our conception of reality. In *Symposium on the Foundations of Modern Physics 1990*, ed. by P. Lahti and P. Mittelstaedt, World Scientific, Singapore, pp. 504–506.

Primas H. (1991): Vor-Urteile in den Naturwissenschaften. In *Wissenschaftstheorie und Wissenschaften*, hrsg. von H. Bouillon und G. Andersson, Duncker & Humblot, Berlin, pp. 49–63.

Primas H. (1991): Reductionism: palaver without precedent. In *The Problem of Reductionism in Science*, ed. by E. Agazzi, Kluwer, Dordrecht, pp. 161–172.

Primas H. (1992): Umdenken in der Naturwissenschaft. *Gaia* **1**, 5–15.

Primas H. (1992): Die Einheit der Wissenschaften: Ein gebrochener Mythos. In *Auf der Suche nach dem ganzheitlichen Augenblick*, hrsg. von C. Thomas, Verlag der Fachvereine, Zürich, pp. 267–271.

Primas H. (1992): Mut zur Ganzheit. 400 Jahre einäugige Wissenschaft sind genug. *Kleine Schriften Nr. 21* (ETH Zürich).

Primas H. (1992): Time-asymmetric phenomena in biology. Complementary exophysical descriptions arising from deterministic quantum endophysics. *Open Systems & Information Dynamics* **1**, 3–34.

Primas H. (1992): Umdenken in der Naturwissenschaft. *Vierteljahrschrift der Naturforschenden Gesellschaft in Zürich* **137**, pp. 41–62.

Primas H. (1992): Warnung! Besprechung von *The Meaning of Quantum Theory* von J. Bagott (Oxford University Press). *Nachrichten aus Chemie, Technik und Laboratorium* **40**, 1152–1153.

Primas H. (1992): Ein Ganzes, das nicht aus Teilen besteht. Komplementarität in den exakten Naturwissenschaften. In *Mannheimer Forum*, Boehringer, Mannheim), pp. 81–111. Reprinted in *Neue Horizonte 92/93*, Piper, München 1993, pp. 81–111.

Primas H. (1992): A propos de la mécanique quantique des systèmes macroscopiques. In *Erwin Schrödinger. Philosophie et Naissance de la Mécanique Quantique*, ed. par M. Bitbol et O. Darrigol, Editions Frontières, Gif-sur-Yvette, pp. 385–402.

Primas H. (1992): Es gibt keine Einsicht ohne innere Bilder. *Gaia* **1**, 311–312.

Primas H. (1993): The Cartesian cut, the Heisenberg cut, and disentangled observers. In *Symposia on the Foundations of Modern Physics 1992*, ed. by K.V. Laurikainen and C. Montonen, World Scientific, Singapore, pp. 245–269.

Primas H. (1993): Mesoscopic quantum mechanics. In *Symposium on the Foundations of Modern Physics 1993*, ed. by P.J. Lahti, P. Busch and P. Mittelstaedt, World Scientific, Singapore, pp. 324–337.

Primas H. (1994): Vom sanften Umgang mit der Natur. In *Denkanstösse '95. Ein Lesebuch aus Philosophie, Natur- und Humanwissenschaften*, hrsg. von H. Bohnet-von der Thüsen, Piper, München, pp. 173–177.

Primas H. (1994): Umdenken in der Naturwissenschaft. *Rotary* **69**(8), 48–51.

Primas H. (1994): Endo- and exo-theories of matter. In *Inside Versus Outside. Endo- and Exo-Concepts of Observation and Knowledge in Physics, Philosophy, and Cognitive Science*, ed. by H. Atmanspacher and G.J. Dalenoort, Springer, Berlin, pp. 163–193.

Primas H. (1994): Realism and quantum mechanics. In *Logic, Methodology and Philosophy of Science IX*, ed. by D. Prawitz, B. Skyrms and D. Westerstahl, Elsevier, Amsterdam, pp. 609–631.

Primas H. (1994): Hierarchic quantum descriptions and their associated ontologies. *Symposium on the Foundations of Quantum Mechanics 1994*, ed. by K.V. Laurikainen, C. Montonen and K. Sunnarborg, Editions Frontières, Gif-sur-Yvette, pp. 210–220.

Atmanspacher H., Primas H., Wertenschlag-Birkhäuser E., Hrsg. (1995): *Der Pauli-Jung-Dialog und seine Bedeutung für die moderne Wissenschaft*, Springer, Berlin.

Primas H. (1995): Über dunkle Aspekte der Naturwissenschaft. In *Der Pauli-Jung-Dialog und seine Bedeutung für die moderne Wissenschaft*, hrsg. von H. Atmanspacher, H. Primas, and E. Wertenschlag-Birkhäuser, Springer, Berlin, pp. 205–238.

Atmanspacher H. and Primas H. (1996): The hidden side of Wolfgang Pauli. An eminent physicist's extraordinary encounter with depth psychology. *Journal of Consciousness Studies* **3**, 112–126. Reprinted in *Journal of Scientific Exploration* **11**, 369–386 (1997).

Primas H. (1996): Synchronizität und Zufall. *Zeitschrift für Grenzgebiete der Psychologie* **38**, 61–91.

Amann A. and Primas H. (1997): What is the referent of a non-pure quantum state? In *Potentiality, Entanglement, and Passion-at-a-Distance. Quantum Mechanical Studies in Honor of Abner Shimony*, ed. by R.S. Cohen, M. Horne, and J. Stachel, Kluwer, Dordrecht, pp. 9–29.

Primas H. (1997): The representation of facts in physical theories. In *Time, Temporality, Now*, ed. by H. Atmanspacher and E. Ruhnau, Springer, Berlin, pp. 243–266.

Primas H. (1998): Emergence in exact natural sciences. *Acta Polytechnica Scandinavica* **Ma 91**, 83–98.

Primas H. (1998): Basic elements and problems of probability theory. *Journal of Scientific Exploration* **13**, 579–613. Primas, H. (1999)

Primas H. (1999): Einige Bemerkungen zum Dialog zwischen Philosophie und Naturwissenschaft. In *Die Aufgaben der Philosophie heute*, hrsg. von V. Hösle, P. Koslowski and R. Schenk, Passagen Verlag, Wien.

Primas H. (2000): Asymptotically disjoint quantum states In *Decoherence: Theoretical, Experimental, and Conceptual Problems*, ed. by P. Blanchard, D. Giulini, E. Joos, C. Kiefer, I.-O. Stamatescu, Springer, Berlin, pp. 161–178.

Atmanspacher H. and Primas H. (2001): Wolfgang Paulis verborgene Seite. Ein aussergewöhnlicher Physiker begegnet der Tiefenpsychologie. *Beiträge zur Integralen Weltsicht* **16**, 19–42.

Primas H. (2002): Hidden determinism, probability, and time's arrow. In *Between Chance and Choice. Interdisciplinary Perspectives on Determinism*, ed. by H. Atmanspacher and R. Bishop, Imprint Academic, Thorverton, pp. 89–113.

Primas H. (2002): Kas keeniat saab taandada füüsikale? *Akadeemia (Eesti Kirjanike Liidu kuukiri Tartus)* **14**(9), 1910–1949.

Primas H. (2002): Fascination and inflation in science. In *The Role of Philosophy of Science and Ethics in University Science Education*, ed. by T.B. Hansen. NSU Press, Göteborg, pp. 72–90.

Atmanspacher H. and Primas H. (2003): Epistemic and ontic quantum realities. In *Time, Quantum and Information*, ed. by L. Castell and O. Ischebeck, Springer, Berlin, pp. 301-321. Reprinted in *Foundations of Probability and Physics-3*, ed. by A. Khrennikov, American Institute of Physics, New York 2005, pp. 49–61.

Primas H. (2003): Time entanglement between mind and matter. *Mind and Matter* **1**, 45–83.

Atmanspacher H. and Primas H. (2006): Pauli's ideas on mind and matter in the context of contemporary science. *Journal of Consciousness Studies* **13**(3), 5–60.

Primas H. (2007): Non-Boolean descriptions for mind-body problems. *Mind and Matter* **5**, 7–44.

Atmanspacher H. and Primas H. (2009): Introduction. In *Recasting Reality. Wolfgang Pauli's Philosophical Ideas and Contemporary Science*, ed. by H. Atmanspacher and H. Primas, Berlin. Springer, pp. 1–10.

Primas H. (2009): Complementarity of mind and matter. In *Recasting Reality. Wolfgang Pauli's Philosophical Ideas and Contemporary Science*, ed. by H. Atmanspacher and H. Primas, Berlin. Springer, pp. 171–209.

Atmanspacher H. and Primas H., eds. (2009): *Recasting Reality. Wolfgang Pauli's Philosophical Ideas and Contemporary Science*, Springer, Berlin.

Primas H. (2016): *Knowledge and Time*, Springer, Berlin (posthumous publication, edited by H. Atmanspacher).

Printed in the United States
By Bookmasters